Early Childhood:

# THE DEVELOPMENT OF SELF-REGULATORY MECHANISMS

# LIST OF CONTRIBUTORS


VICTOR H. DENENBERG

BÄRBEL INHELDER

ROBERT M. KRAUSS

ERIC H. LENNEBERG

JOHN C. LOEHLIN

ROBERT B. McCALL

DONALD L. PETERS

MARK R. ROSENZWEIG

ALFRED STEINSCHNEIDER

M. X. ZARROW

EARLY CHILDHOOD

# The Development of Self-Regulatory Mechanisms

*DWAIN N. WALCHER and DONALD L. PETERS*

The Pennsylvania State University
University Park, Pennsylvania

ACADEMIC PRESS New York and London 1971

ACADEMIC PRESS, INC.
111 Fifth Avenue, New York, New York 10003

*United Kingdom Edition published by*
ACADEMIC PRESS, INC. (LONDON) LTD.
Berkeley Square House, London W1X 6BA

LIBRARY OF CONGRESS CATALOG CARD NUMBER: 78-137601

PRINTED IN THE UNITED STATES OF AMERICA

*"Our social philosophy in the United States is built upon the notion of self regulation; hence it is crucial that we increase our understanding of how self regulation can be fostered in individuals."*

Donald H. Ford, *Dean*

*College of Human Development*
*The Pennsylvania State University*

## CONTENTS

*Contributors* xi
*Conference Participants* xiii
*Foreword* xv
*Acknowledgment* xix

**Conference Overview**

Biological Regulatory Mechanisms 2
Early Self-Regulatory Behavior 5
Thoughtful Self-Regulation 8
Models of Human Self-Regulation 11
Implications for Future Action 12

**Role of Experience in Development of Neurophysiological Regulatory Mechanisms and in Organization of the Brain**
*Mark R. Rosenzweig*

Trial-and-Error Self-Organization as a Supplement to Limited Genetic Information 16
Ontogeny of Regulation of Intake of Food 17
Measurable Changes in Brain as Consequences of Differential Experience 24
Discussion 28
Conclusions 30
References 31
Discussion 34

**Effects of Handling in Infancy upon Adult Behavior and Adrenocortical Activity: Suggestions for a Neuroendocrine Mechanism**
*Victor H. Denenberg and M. X. Zarrow*

Effects of Handling in Infancy upon Later Behavior 40

Effects of Handling in Infancy upon the Adrenal and the Plasma Corticosterone Response and Emotionality 43
Neonatal Sex Hormones and Behavior: A Model for Infantile Stimulation 48
Studies Involving Corticosterone as Part of the Mechanism of Infantile Stimulation 49
Studies on the Pituitary-Adrenal System of the Neonatal Rat 58
Summary 62
References 62
Discussion 64
Remarks of Dr. Lewis Lipsitt 68

**Determinants of an Infant's Cardiac Response to Stimulation**
*Alfred Steinschneider*

General Procedure 75
Response Measures 76
Average Response 81
Monophasic or Diphasic Response 84
Initial Value 86
Acceleration or Deceleration (Crossover Point) 89
Behavioral State 92
General Summary 98
References 100
Discussion 102

**Attention in the Infant: Avenue to the Study of Cognitive Development**
*Robert B. McCall*

The Measurement of Attention 109
Complexity, Contour, Rate of Change, and Age 111
Response to Discrepancy 115
Conclusions 131
References 133
Discussion 137

**The Sensory-Motor Origins of Knowledge**
*Bärbel Inhelder*

The Development of the Universe: The Origin of the Categories of Knowledge 142
From Sensory-Motor Activities to the Semiotic Function 148
Conclusions 154
References 155

**Of Language Knowledge, Apes, and Brains**
*Eric H. Lenneberg*

Language Knowledge: Formal Constants of Cognitive Achievement 157
Apes: Heuristics of Cross-Species Comparison of Cognitive Function 170
Brains: Preliminaries to Theorizing 175
Discussion 180
References 183
Discussion 184

**The Interpersonal Regulation of Behavior**
*Robert M. Krauss*

Text 187
References 202
Discussion 203

**Regulatory Functions In Computer Models**
*John C. Loehlin* 209

Computer Program Structures 210
Examples of Self-Regulation 213
Final Comments 225
References 225
Discussion 226

**The Development of Self-Regulatory Mechanisms: Epilog**
*Donald L. Peters*

Self-Regulatory Mechanisms 232
The Dynamic Interactional View of Human Development 234
The Processes of Individualization 235
Research Implications 236
References 238

*Author Index* 241
*Subject Index* 246

## CONTRIBUTORS

Numbers in parentheses indicate the pages on which the authors' contributions begin.

VICTOR H. DENENBERG (39), Professor of Bio-Behavioral Sciences and Psychology, University of Connecticut, Storrs, Connecticut

BÄRBEL INHELDER (141), Professor of Psychology, University of Geneva, Geneva, Switzerland

ROBERT M. KRAUSS (187), Professor of Psychology, Columbia University, New York, New York

ERIC H. LENNEBERG (157), Professor of Psychology, Cornell University, Ithaca, New York

JOHN C. LOEHLIN (209), Professor of Psychology, University of Texas at Austin, Austin, Texas

ROBERT B. McCALL (107), Chairman and Senior Investigator, The Fels Research Institute, Yellow Springs, Ohio

DONALD L. PETERS (231), Professor of Human Development, The Pennsylvania State University, University Park, Pennsylvania

MARK R. ROSENZWEIG (15), Professor of Psychology, University of California, Berkeley, California

ALFRED STEINSCHNEIDER (73), Professor of Pediatrics, State University of New York, Syracuse, New York

M. X. ZARROW (39), Professor of Bio-Behavioral Sciences and Psychology, University of Connecticut, Storrs, Connecticut

## CONFERENCE PARTICIPANTS

George A. Borden
*Associate Professor of Speech*
*The Pennsylvania State University*

Vladimir de Lissovoy
*Associate Professor of Child Development & Family Relationships*
*The Pennsylvania State University*

Victor H. Denenberg
*Professor of Bio-Behavioral Sciences and Psychology*
*The University of Connecticut*

David Elkind
*Professor of Psychology*
*The University of Rochester*

Louise Guerney
*Individual and Family Consultation Center*
*The Pennsylvania State University*

Helen A. Guthrie
*Associate Professor of Foods and Nutrition*
*The Pennsylvania State University*

Dale B. Harris
*Professor of Psychology and Human Development*
*The Pennsylvania State University*

David Hines
*Department of Behavioral Science*
*The Milton S. Hershey Medical Center*

Bärbel Inhelder
*Professor of Psychology*
*University of Geneva (Switzerland)*

Robert M. Krauss
*Professor of Psychology*
*Columbia University*

Eric H. Lenneberg
*Professor of Psychology*
*Cornell University*

Lewis Lipsitt
*Professor of Psychology*
*Director, Child Study Center*
*Brown University*

John C. Loehlin
*Professor of Psychology*
*University of Texas (Austin)*

Eleanor Maccoby
*Professor of Psychology*
*Stanford University*

Robert B. McCall
*Chairman and Senior Investigator*
*The Fels Research Institute*

Evan G. Pattishall, Jr.
*Professor and Chairman of Behavioral Science*
*The Milton S. Hershey Medical Center*

Donald L. Peters
*Professor of Human Development*
*The Pennsylvania State University*

Ruth L. Pike
*Professor of Nutrition*
*The Pennsylvania State University*

Mae Rosenberg
*Program Analyst, Office of Program Planning and Evaluation*
*The National Institute of Child Health and Human Development*

Mark R. Rosenzweig
*Professor of Psychology*
*The University of California (Berkeley)*

Helen I. Snyder
*Associate Professor of Educational Psychology*
*The Pennsylvania State University*

Aletha Stein
*Assistant Professor of Human Development*
*The Pennsylvania State University*

Alfred Steinschneider
*Professor of Pediatrics*
*State University of New York (Syracuse)*

Raymond G. Studer
*Professor and Director, Division of Man-Environment Relations*
*The Pennsylvania State University*

Hoben Thomas
*Assistant Professor of Psychology*
*The Pennsylvania State University*

Dwain N. Walcher
*Director, The Institute for the Study of Human Development*
*The Pennsylvania State University*

Elizabeth R. Craft
*Stenotypist*
*Washington, D. C.*

Leora W. Wells
*Science Writer*
*Springfield, Virginia*

# FOREWORD

The four day conference on Early Childhood: The Development of Self-Regulatory Mechanisms, held at The Pennsylvania State University in January, 1970, was one of the first attempts to assemble some of the scattered knowledge about self-regulatory mechanisms as an aspect of the development of living organisms. The principal purpose of the conference was to create a working foundation for bridging the gap between analytic and social concerns.

The synthesizing role is the key concept around which the Institute for the Study of Human Development was organized. This Institute, one of five units in the University's College of Human Development, was conceived as a device to facilitate multidisciplinary examination of problem-focused issues.

Scientific description of man emerges from many fields, thus, when a conference focuses upon the development of one aspect of human behavior, it is well to temper "great expectations" with the realities of academic interchange of ideas. No conclusions were expected and the "funding fathers" had as a primary purpose the establishment of a setting to enable outstanding scholars to meet for a free interchange of ideas.

To permit as much informal discussion as possible, the number of participants was kept small. Each visiting scientist represented a different focus of research interests. Seven formal presentations by visiting scientists were scheduled. Since one of the speakers was ill and unable to attend, her paper was distributed to the participants, who then received the unexpected bonus of an eighth thought-provoking presentation. Four additional visiting scholars complemented the formal presentations by questions, interpolations, and manifestos of their own.

Sharing responsibility for the success of the conference were the members of The Pennsylvania State University faculty who were active participants in the group. Each visiting scientist had a counterpart faculty host with whom he shared professional interests. Although members of this university faculty did not make formal presentations, they and the other visiting scientists raised many stimulating questions in response to the presentations of the speakers and contributed pertinent data from their own and other studies. A scientific writer served to integrate and to interpret the language of specialized research.

Graduate students were invited to sit in the observers' gallery during the conference sessions and to meet with the participants at lunch and coffee breaks. This served the double purpose of enabling the students to meet and talk with scientists with whose work they were familiar and enabling the participants to see their own work through fresh eyes as a result of the questions and comments of the students.

The conference was under the general chairmanship of Dr. Dwain N. Walcher, Director of the Institute for the Study of Human Development. Dr. Donald L. Peters and Dr. Helen A. Guthrie of the College of Human Development served as scientific chairman and co-chairman.

That the noble purpose of the conference was attained in some degree is attested by the fact that two scientists, working for fifteen years in the same field, met face to face for the first time. What latent functions were accomplished time alone will tell. It is hoped that research interests have been served and questions will continue to be generated.

It soon became clear, however, that self-regulatory mechanisms as an encompassing concept were not always the central issue of the discussion. The living organism as an open system seldom can be probed in one area without response at its interfaces. When the system is examined in a developmental perspective, the dynamics of growth, maturation, and learning defy clearcut conceptualization, to say nothing of research findings. In fairness to the distinguished participants and to the readers of this tome, it would be better to cast this conference as one oriented to an exploration of mechanisms propaedeutic to the development of self-regulatory behavior.

Regulation has a wide range of meanings. It implies order and adaptation. Regulatory devices are prevalent in simple as well as in sophisticated mechanical and electronic mechanisms. Usually these are external regulators which guard parameters of a given function or serve to initiate some type of adjustive reaction such as automatic correction for drift or the breaking of a circuit in an overload. In recent years highly sophisticated electronic systems have been developed with intricate capabilities of "self-regulation." Regulatory functions in living organisms take many forms. The single cell in a chemical transaction with its environment manifests order of function. The complex patterns of migrations of particular species noted as "instinctual" behaviors are as yet unexplained regulatory mechanisms. Still at anotner level the internalization of value systems or propensities for given behaviors is illustrative of a type of human self-regulation.

Man as a member of society is bound by limitations and possibilities. Human nature and social order represent vast parameters of interrelationships manifesting endless questions in regard to structure and function of each. To be human is to be comprised of systems and subsystems each with its own regulatory mechanism. To be a member of society is to be regulated and, eventually, to assume a degree of self-regulation. The development of regulatory behavior has its basic roots in the organismic nature of man.

The concept of organismic regulation leads to description of behavior of organisms in the process of self-maintenance; it is progressive readjustment of a dynamic system disturbed by changes incidental to processes of growth and metabolism and by environmental alterations. In man the action system is understood only as the totality of organic function is considered. The elaborate physiological mechanisms, highly differentiated, specialized in structure and function, and coordinated through chemical processes and a highly developed neural network, permit a wide range of adaptive behaviors.

In the course of experiments with protozoa, M. S. Jennings came to view behavior as "action systems" of organisms in process of self-maintenance, a progressive readjustment of dynamic conditions disturbed in the process of growth and metabolism or by alterations in the environmental system. Behavior was seen as changed in response to conditions. Jennings recognized that internal processes of organisms could serve to elicit change of behavior, but always in response to conditions.

The maintenance of steady physical and chemical states within an organism is basic to the life process. Recognizing the complexity of physiological processes in the intricate subsystems of the human organism Cannon suggested a special designation for these states to differentiate them from the notion of immobility or stagnation. "Homeostasis" as a regulatory process was postulated to avoid the implication of a static state and to suggest ". . . a condition which may vary, but which is relatively constant."

In no living organism are self-regulating systems so evident as they are in man. Through the span of development a series of transitions require adaptation to internal changes of growth and maturation as well as to the demands of the environment. A variety of messages are transmitted through physical and chemical components, and neural systems involving intricate feedbacks by which the organism develops its self-regulation, self-repair, and self-direction. The organism is an open system with continual inputs and outputs. Each individual is unique in hereditary aspects and specific experiences. Thus the focus of inquiry must extend beyond the measure of observable product to include the question of processes.

The study of processes requires an approach not bound by traditional assumptions of linear relationships. The organism as an active energy system selects, rejects, incorporates, or initiates responses with circular or reciprocal feedbacks that can defy a clear antecedent definition of the traditional stimulus response model. This can be seen in much developmental research which implies behavior-personality to be a direct result of an interaction between the organism and the environment. Such a model does not take into account the fact that the organism is never in a steady state. In fact, the dynamic alterations of the multiple systems in a developing organism are "intervening variables" which mediate an observable reaction.

The complexities of dealing with dynamic states of organisms encourage inquiry limited to experiments oriented to elegant design and to measurement of relationships among variables. These models have been productive in the creation of a solid base of research techniques and a vast body of knowledge. It is clear, however, that a great need exists for a systematic study of organisms as ongoing, functioning, emerging totalities. This view has been classically promulgated by Von Bertalanffy.

Dramatic progress has been made in the application of self-regulating devices in the technological sciences. From the safety valve, thermostat, and governor, increasing scientific sophistication has produced servomechanisms, automation, and cybernetics. The intricacies of feedback loops, storage systems, and "self-generating" processes have aroused great interest in the application of this knowledge to the study of living organisms. Unquestionably this area of the science holds much promise for the understanding of man.

This conference offers many points of view. The reader may be attracted or dismayed by the variety of research questions and the diversity of techniques. A careful appraisal of the themes underlying the presentations, discussion, and summation will reveal much common substance. Most systems are not immediate in their response to feedback. They tend to lag a bit and this characteristic is noted as "hunting"; perhaps this was the essence of our meeting.

Vladimir deLissovoy, Ph.D.

*College of Human Development*
*The Pennsylvania State University*

## ACKNOWLEDGMENT

Mrs. Leora Wood Wells served as science writer for this conference. Her outstanding skill in handling interdisciplinary scientific material has made the overview and the discussions sections of this publication more meaningful to those not in attendance.

# CONFERENCE OVERVIEW

The conference theme of the development of regulatory mechanisms in early childhood was subdivided into four central areas of interest:

Biological Regulatory Mechanisms
Early Self-Regulatory Behavior
Thoughtful Self-Regulation
Models of Human Self-Regulation

Inevitably, this division was somewhat arbitrary, since biology, cognitive development, various aspects of behavior, and the influences of environment and other external stimuli are inseparably intertwined in human development. However, these designations served as rough guidelines to enable the participants to sort out the points at which their particular research interests converged with the central theme.

The presentations were set within the broad frameworks of animal studies, computer models of regulatory functions, and studies of human development from the preconceptual level throughout a range of influences spanning, in some cases, several generations. Although it might seem a long step from measures of emotionality in rats to fear and anger reactions in computer programs or impulse control in children, many possible relationships among such phenomena became apparent as the conference proceeded. Time limitations prevented these relationships from being spelled out and discussed in detail, but the groundwork was laid for productive interchanges of ideas in the future. Before the conference ended, plans for joint research had already been laid by at least two of the participants and members of the group urgently requested that they be invited to reassemble at intervals to continue what had been begun at the conference.

The first two speakers dealt with neurophysiological and neuroendocrinological regulatory mechanisms. This was followed by a discussion of the infant's cardiac response and then a discussion of his attention patterns as avenues for the study of cognitive development. The next three papers, which focused around the area of thoughtful self-regulation, covered the sensory-motor origins of knowledge, the nature of language knowledge, and factors affecting ability to communicate. The final speaker described some of the self-regulatory features of computer programs which are similar to human regulatory mechanisms. At the final luncheon session of the conference, Dean Donald H. Ford of the College of Human Development keyed in the conference discussions with the overall goals of the College and the University, and the participants looked ahead to some of the developments which might be expected to grow out of the exchange of ideas that had taken place during the conference.

The papers of the visiting scientists are given in full in this volume, together with summaries of the discussion which followed each presentation and lists of references to related studies. What follows here is not a summary of the papers and discussions but a selection of some of their highlights which relate most closely to the conference theme.

## BIOLOGICAL REGULATORY MECHANISMS

Many studies done with rats and other mammals have provided important clues to the way neurophysiological self-regulatory mechanisms develop. However, much additional research is needed to test out various hypotheses.

For example, it appears that some structural and behavioral regulation is acquired through a trial and error process of self-organization. The play instinct of young mammals represents such a process. Play activities serve the utilitarian purpose of helping the young animal develop coordination and move toward optimal organization of its behavior through trial and error experience.

Why is this necessary? According to some investigators, organisms do not contain enough genetic information to specify either bodily structure or behavior in full detail. The genetic information contained in organisms is sufficient only to provide a "master structure" in the brain, they say, and the details are organized through interaction with the environment. Other investigators, however, feel that certain detailed organizational and behavioral capacities are present in mammals at birth but that these degenerate and become disorganized if the young animals are kept in isolated or unstimulating environments.

The control of food intake in rat pups is one example of organization present at birth. The young animal is highly selective, accepting only the types of food its body is able to assimilate at each stage of development. It has been shown that interference with this organizational sequence—for example, through premature weaning—often has deleterious effects on later ability to form conditioned responses, but that these effects can be altered or prevented by adjusting certain nutritional and social factors.

The role of learning has received insufficient attention in most studies of the ontogeny of regulation of food intake, even though it has been demonstrated that very young animals quickly learn to change their behavior if the situation involves activities which must be present or develop very early in their lives. For example, puppies and kittens must suckle in order to survive. Consequently, investigators have been able to induce sucking as a conditioned response to certain stimuli within the first few days of life; whereas conditioned withdrawal to shock or tactile stimulation cannot be demonstrated before the second or third week of age.

Variations in rearing conditions of rat pups during the first few days of life have been shown to affect their regulation of food intake. Under normal rearing conditions, an infant rat will suckle a lactating female. An incubator-raised, tube-fed rat pup, however, will seek the familiar food catheter even when a lactating female rat is accessible to him.

Studies with both postweaning and preweaning rats have shown that differential experience has significant effects on the brain. Measurable changes have been observed in brain weight, brain chemistry, and brain anatomy as a result of rearing in enriched or impoverished conditions. Some evidence suggests that neural circuits of the brain are altered as a result of differential experience, but this has not yet been clearly demonstrated. It is difficult at this stage of research to assess what these brain changes mean. Some may reflect the development of self-regulatory mechanisms, and other processes of learning and memory storage, but these are simply hypotheses which need to be tested by much further research.

Findings of recent studies suggest a number of important possibilities in relation to neuroendocrine regulatory mechanisms. Handling in infancy appears to affect both the adult behavior of rats and their adreno-cortical activity. In development, at least two major sets of processes are at work:

i. A biochemical set of events which is imbedded around the hypothalamic region and is involved with affective actions, and
ii. A set of events related to cognitive development.

Studies have shown that stimulation of rat pups in early infancy has profound effects upon affective behaviors and very little effect upon cognitive behavior, at least as measured by performance on the Hebb–Williams maze, even though handling in infancy makes the rat pup a great stimulus-seeker and more interested in manipulating novel objects.

Empirically, it appears that the best age to manipulate a rat to produce a better cognitive, perceptual problem-solving animal is after its eyes are open at about 14 days. If the investigator wishes to change the rat's physiological capabilities for emotional behavior, the best age is within the first 5 or 10 days of life.

Some investigators feel that the parallel in the human is that manipulation during the first 3 to 6 months of life will produce affective results while manipulation at later ages produces more cognitive effects. It may be, these researchers say, that the neural systems are insufficiently developed for longterm cognitive retention at the younger age levels.

Other investigators feel that this is true only of avoidance conditioning and that cognitive effects can be obtained earlier if the experiments are related to the stage of the individual's development. In children, they say, the potentials for both cognitive and affective functioning appear to be present at an early age, but they remain so closely bound together that it is difficult to differentiate them until about the age of 6 or 7.

In relation to emotionality, specific effects of handling have been demonstrated with rats. When defecation and activity in the field are used as measures, animals which have been handled are behaviorally less emotional than nonhandled animals. The more intense the stimulation, or the greater the magnitude of the stimulation in infancy, the less emotional the animal is in later life. These less emotional animals also are less reactive physiologically in terms of adrenal corticosterone response.

In exploring the significance of this response, researchers turned to what has been observed about the effects of sex hormones. They noted that it is possible, up to a certain point, for the brain to develop either toward a male orientation or toward a female orientation. The way this takes place is one of the best examples of brain organization we have. The sex hormones that are present in a prenatal or newborn animal are responsible for organizing the brain of that animal as male or female, while the function of sex hormones in adulthood is to activate an already organized set of behaviors and biological systems.

Corticosterone is present in newborn rats. The investigators hypothesized that corticosterone may be secreted in the infant animal the same way testosterone and estrogens are secreted and that it may act upon the brain to modify emotionality. It is released as a function of certain types of strong stimulation, and also by the mild stimulus of handling.

There are two forms of stimuli which impinge upon an animal: one set goes through the central nervous system and affects it by nervous system outlets, the other set acts systemically, directly on the blood stream. Corticosterone may be not only part of the response system affected by handling but also may be part of the mediating central mechanism.

Research has demonstrated that manipulation of a pregnant female rat brings out emotional differences in her offspring. When corticosterone is injected into a pregnant rat, it crosses the placental barrier. By the time the rat pup is 2 days old, corticosterone can be identified in the hypothalamus of the offspring. On the other hand, the amount of corticosterone ingested by the rat pup through the mother's milk is so small that it is very unlikely to be biologically meaningful.

Several nongenetic effects on young rats have been demonstrated. In their field activity, offspring of nonhandled mothers behave like nonhandled females even when reared by handled mothers. Effects span more than one generation. A study of preweaning and postweaning activities of rat pups indicated that the behavior of the young rat is determined in part not only by what happened to its mother during her early life but also by what happened to the animal's grandmother when she was an infant.

These findings appear to tie in with studies of human nutritional status which show that the nature and character of infants relate clearly to the mother's nutritional status at puberty, long before she begins to reproduce. Interlinking the findings of such studies could point to directions for further exploration of how the various qualities of offspring can be shaped by preconceptual influences on their forebears.

Extreme caution should be exercised in extrapolating the findings of animal studies to humans. At present we can only say that certain factors produce clear effects in certain experimental animals, and we ought, therefore, to consider what related experiments would be worth doing in humans.

However, we have been able to learn much of what we know about human infant behavior because of the groundwork laid by animal studies. Many advances in the field of child development have come about because techniques have been worked out in animal laboratories, which it has been possible later to apply in human studies. This is not, of course, true of all research problems. There are also a number of phenomena and techniques which can be studied and applied in the animal laboratory which cannot be studied in humans; in these cases, one must rest content with the animal data. With respect to certain other types of phenomena, it is possible to go directly to human research without prior animal studies.

## EARLY SELF-REGULATORY BEHAVIOR

Neonates show many physiological responses to environmental stimuli. Although these responses show certain predictable patterns, the reactions of each infant within the general pattern are individually characteristic.

Cardiac rate is one of the easiest autonomic responses to measure, and studies have shown that both acceleration and deceleration reflect sensitivity to external stimuli. The cardiac deceleration in the waking state of the neonate which occurs in response to stimulation is thought to be a component of the orientation reflex. Stimuli during sleep are most apt to cause cardiac acceleration. This appears to be a component of the defense reaction, a set of responses which are associated with efforts to limit the effects of subsequent stimulation. The magnitude, form, and direction of these responses do not depend solely upon the stimulus conditions but depend also upon organismic variables, developmental factors, and the behavioral state of the infant.

Stimuli used in studies of neonates and infants have included white noise and jets of air on the skin. Although individual differences in response are noted, in general, stimulation of the newborn with a 5 second air stream produces a steady increase in heart rate which achieves its peak some seconds after the onset of stimulation and then decreases to a level approaching or less than that prior to stimulation. When white noise is used as the stimulus, increasing the intensity results in greater peak magnitudes and shorter amounts of time required to reach these magnitudes. Although neonates show different degrees of response to stimulation, each shows a certain consistency in his responses; for example, a neonate who shows a large increase in cardiac rate in response to weak sounds also shows a large increase in response to loud sounds.

The response of older children to the white noise stimulus differs from that of neonates. Instead of an initial increase, the older children demonstrate an initial decrease in heart rate. This is followed by acceleration but the rate remains below prestimulus level. This is true even when the same children are tested as infants and again at older ages.

The prestimulus level of organic function is a major determinant of cardiac response to stimulation. On the average, prestimulus heart rate is negatively correlated with peak magnitude. Temporal response measures, however, have not been found to be a function of the prestimulus heart rate.

To understand cardiac response to stimulation, it is important to consider both the initial value and the particular variables which influence the extent to which the initial value determines response. Most babies have a specific initial heart rate value at which they give no heart rate change with stimulation. If stimulation occurs at a time when the prestimulus heart rate level is lower than this point, the heart rate will show an initial increase. If the stimulus occurs when the prestimulus heart rate is greater than the crossover point, the heart rate will show a decelerative response. Neither stimulus duration nor the age of the child affects these values, but stimulus intensity is a significant variable for both newborns and older children.

Thus it is clear that the direction of the cardiac rate response to stimulation is dependent upon at least two variables, the prestimulus heart rate and the intensity of stimulation. In the majority of published studies, the average cardiac rate response of newborns is described to be primarily one of acceleration. However, the newborn can and will respond to stimulation with deceleration if certain test conditions prevail.

The unstimulated cardiac rate is correlated with the infant's behavioral state, tending to be lower when he is asleep than when he is awake. However, the exact effects of behavioral state on the magnitude of the accelerative or decelerative response have proved difficult to measure, and investigators have reported diametrically opposite results.

Studies suggest that the cardiac response in a neonate forms a continuum of response shapes influenced by the prestimulus level and anchored at one end by an accelerative curve and at the other end by a decelerative curve, with a diphasic curve in between. Some of the conditions predisposing to development of the various types of responses within this continuum can be specified. The influence of both stimulus intensity and age can be viewed within this conceptual scale.

A number of studies have attempted to determine which components of the response curve best fit the orientation or defense reaction patterns. Data from these studies can be viewed only as indirect evidence of the functional implications of cardiac rate changes. Further studies are needed to clarify the relationship between cardiac rate change and sensitivity to stimulation as shown in the deceleration and acceleration responses. Studies are also needed to explore many other variables such as the influences of sex, various personality or behavioral characteristics, age-related physiological characteristics, and the effects of orienting and defense reflexes on learning.

Studies of attention in infants provide an interesting avenue to the study of cognitive development. Since the selection of stimuli toward which to orient oneself is one of the more important features of self-regulation, it is useful to know what determines an infant's choice of what to look at.

There are many variables that may make a stimulus attractive or unattractive. Discrepancy is one of these, and attentional preference for novel stimuli as opposed to familiar stimuli has been shown to appear as early as 4 weeks of age. A stimulus tends to lose its potency as the infant becomes familiar with it and learns from it. This progressive decline in attention may reflect the process of acquiring a memory for the stimulus.

Attention is a cognitive process which acts as a filter on the environment. The stimulus that passes through the filter may be encoded, remembered, and used to guide future behavior, thus affecting the infant's ability to interact with the environment.

What captures an infant's attention is related to what he is able to assimilate at any given time. It appears that younger infants give more attention to visually simple stimuli and more mature infants give greater attention to more complex stimuli. Although there have been several studies which indicate that infants attend to and process increasingly complex stimuli as they mature, the mechanism that mediates these changes is still not known.

Attention is a cluster concept with many response facets. Some children respond with differential looking, cardiac, pupillary, or other response variables to auditory and visual events.

One way to measure how much attention an infant gives a stimulus is to observe how long he looks at it. However, the length of time an infant looks at a stimulus may mean a number of different things. It may reflect the magnitude of the discrepancy in the stimulus, or how much prior familiarization to it the child has had. It may reflect the speed with which the infant assimiliates what he sees and habituates to it, or it may reflect the fact that he is by nature a "short looker" or a "long looker." It may mean that the infant is studying the stimulus in depth and continuing to take in information about it. On the other hand, prolonged looking may sometimes be related to a lack of information processing. Additional information must be obtained before the meaning of these several responses is clear.

The way a child habituates in relation to his first fixation on a visual stimulus is predictive of his response to discrepancy. Rapid habituators and short lookers respond more to discrepancy than do slow habituators or long lookers. Rapid habituators show more cardiac response to discrepant stimuli than slow habituators.

It is evident that infants regulate their attentional behavior in response to stimuli in a number of ways. Better theories and more data are needed on the meaning of these various responses and the dynamics that produce differences in response. Such assessments might provide more salient predictors of later intellectual performance than do infant intelligence tests now available.

## THOUGHTFUL SELF-REGULATION

Regulatory systems are found on all levels of the organism's functioning. They constitute one of the most general characteristics of life and the most general mechanisms to be found in both organic and cognitive behaviors.

Indeed, all development during the sensorimotor period may be conceived in terms of self-regulation. Each self-regulatory system presupposes the existence of several types of feedback resulting from the interaction of the child and the environment and from changes in the environment. One type of feedback results from coordination of the individual's general activities. The other results from the progressive differentiation between the individual's activities and the objects in his environment.

During his first 6 or 8 months, the child is capable of following the movement of objects, but when they disappear from view he does not credit them with the power to continue their course. He continues to seek for them where they had been.

Around the middle of the second year, however, the infant begins to make intelligent connections between what happens and how this is achieved. He assimilates new data into schemes on which action systems are based. His progress toward elaboration of new behaviors occurs in two successive stages. The first is the extension of what is already known. If the baby discovers that one toy makes a noise when he shakes it, he will shake others as well. At this stage, goal and means are relatively undifferentiated. At the next stage, the baby becomes somewhat more sophisticated, assigning a specific goal to his act before he carries it out.

At this same time, symbolic representation begins to emerge. As the child develops, he moves through progressively more complex levels of sensorimotor behaviors. These serve as regulatory mechanisms which enable him to make his actions converge with symbolic functions so that thought operations develop and he achieves a command of language. Language is a particularly important aspect of a more general cognitive function which is prepared for by the sensorimotor forms of behavioral development which precede it.

It is difficult to define the exact nature of language knowledge. It is not just communication. Language is one means of communication, but to use it for this purpose, we must have a base of language knowledge which includes a grasp of certain orderly relationships, principles, classes, and dynamic processes.

Language knowledge is not the same as its use. In studies of child language, too much attention is devoted to what children say instead of to what they know. We cannot assume that a person lacks language knowledge just because he is nonverbal. Some people can comprehend fully what is said to them and can follow directions, yet cannot respond verbally. Others can repeat verbatim what they hear, yet are unable to use language themselves in a meaningful way.

Controversy rages over whether or not animals have language knowledge. Some animals respond in ways that seem to their human observers to indicate an understanding of language, but it is difficult to sort out how much of this apparent knowledge rests on the observer's interpretation of what goes on in the animal's mind. The criteria used in various studies are often imprecise, irrelevant, or subjective and represent more of a projection of the experimenter's assumptions than knowledge in the animal. However, valid methodologies for such studies can be developed. It is important that this be done, because animal experiments can be helpful in identifying how the brain works in relation to language.

Language tends to be poorly taught because we know too little about which relationships develop first. What can be said specifically about the brain mechanisms involved in language knowledge is still very vague. We cannot tidily locate the site of language knowledge as being in a particular area of the brain. Since language involves active relationships, it follows that the place to look for a neurological basis for language would not be in the structure of the brain, as some investigators have assumed, but in its activities. We cannot map behaviors to structures on a one-to-one basis. We need to study what activities take place at the same time in the visual and auditory areas of the brain and discover what clues they can provide to the question of why man can speak.

We also need to identify what enables man to use this speech or language ability meaningfully. In order to comprehend and produce sentences, the user must possess a certain competence which grows out of the complex, abstract knowledge that underlies all language behavior.

Many psychologists interested in language development focus on investigating the process by which the child extracts from samples of speech to which he is exposed the latent structure from which other sentences can be generated. To the social psychologist, however, the principal focus of interest is more apt to be the way language functions as a means of communication.

Young children talk incessantly, but they rarely place themselves at the point of view of their listeners. They speak as if they are thinking out loud. Experiments have shown marked differences in communication effectiveness as a function of age. This appears to be largely attributable to inadequacies in the speakers' abilities to encode what they say in terms that will be meaningful to the listener.

Age also influences the way children try a second time to communicate if they have not succeeded in the first attempt. Young children simply repeat what they said before without modifying the content. Older children change or embellish their messages, although their new description is not always clearer than the initial one.

Deficient communication can result from either of two causes. Either the speaker does not have adequately communicative messages in his vocabulary repertoire, or he selects a term that is not meaningful to his listener. To communicate effectively, the speaker must not only have a sufficiently large repertoire of concepts; he must also have the ability to select those with potential social meaning. One index of social meaningfulness is communality. When children of kindergarten age are asked to give names to certain abstract shapes, they often respond idiosyncratically with names meaningful only to themselves. Older children give names with greater communality. As a result, their ability to communicate information about the stimuli through language is greater than that of younger children.

In adults, the level of communication may depend upon the target of their messages. In one experiment the subjects were asked to name color chips for later self-identification and for later identification by others. When retested several weeks later, the subjects were able to match up the color chips more easily with the nonsocially encoded names they had given them than with the socially encoded names given them by others. They had difficulty identifying which colors related to which of the nonsocially encoded names given them by other subjects.

In the everyday use of language, we do not yet know what the learning process is through which a person develops the ability to regulate his communication in response to the social needs of the listener. We do know that his expectation of the listener affects the way he talks to him. If he thinks his listener is smart, he will speak to him one way; if he thinks he is stupid, he will address him another way. People use different approaches in talking to children, to adults, to foreigners, to strangers. It would be useful to try to identify what the way people modify their descriptions can tell us about language as an internal, self-regulatory mechanism as well as a regulatory mechanism in social communication.

## MODELS OF HUMAN SELF-REGULATION

Computer programs can exhibit complex and interesting behaviors resembling human behaviors. Most programs contain both cognitive and motivational systems, and these involve a number of self-regulatory mechanisms. Such programs can, therefore, be designed to test out theories concerning human self-regulatory processes. In computer programs, these self-regulatory mechanisms draw upon the model's past experience to govern the behavior, and the model shows a tendency to stability of behavior over time.

Computer programs can be designed to develop emotional attitudes such as love, fear, and anger toward certain objects or situations, and they can be designed to modify these attitudes or take actions based upon them. Computers can be programmed to represent neurotic personality processes in response to conflicts between strongly charged beliefs and to respond to "therapy" which redirects the attention of the program. Two program models can be caused to interact with each other, and the resulting modifications in their attitudes can be studied.

It is possible to build instructions into computer programs which cause them to make decisions on the basis of specified criteria when the appropriate moment arrives. Several simple decisions can be linked together to yield a complex decision, since computer programs have not only a major executive program but many subroutines on which the main program relies for detailed operations essential to achievement of its assignment. They may contain a control principle which enables a high priority operation to interrupt an ongoing process of lower priority.

Computers can be programmed to perform complex intellectual tasks, considering many more possibilities in a situation than human beings are able to do. In problem solving situations, they employ such regulatory mechanisms as adjusting their own evaluations on the basis of their successes or failures and possible outcomes that become apparent.

Robots can be programmed to protect themselves against physical damage by stopping when they encounter certain types of obstacles and thinking about what path they can safely take to get to their destination. This carries interesting parallels to problems of impulse control in children. A curiosity motive or exploratory mode can be built into a robot that will cause the robot to investigate its environment when it has no other explicitly assigned tasks to perform.

Currently a computer program has been developed to represent learning at the level of a human infant just discovering how to manipulate his environment. Although the biological input and output system of the human is much more complex than that of the model, the mechanical system can provide insights into decision-making structures.

If an experimenter wants to examine a whole set of processes that occur within the organism, the computer program may be able to do a better job than the human of looking at variables and arriving at predictions. Although such programs do not provide empirical information about how people do things, they do indicate the implications of whatever assumptions are built into the program. The reliability of their predictive simulation is limited by the fact that there is so much that goes on inside the human organism we do not understand and therefore cannot program into the computer. However, many complicated configurations of ideas can be explored.

Our knowledge of self-regulatory and self-evaluation processes and impulse control in children is very limited at present. Computer programming has provocative implications for studying these processes in greater detail.

## IMPLICATIONS FOR FUTURE ACTION

The most important impact of the conference was the opening of new channels of thinking for the participants, both in relation to their own work and in relation to that of the other scientists who were present. The concept of self-regulation was new to some of them. Some had come to the conference unsure of what it meant and unconvinced that their own work related in any way to the concept. By the time they left, several could see previously unrecognized implications in research already completed and new avenues they might profitably explore in future studies.

The need to encourage both basic and applied research was apparent. The components of Human Development as a discipline, the basic research to be done, and the relevance of this to immediate situations all came to have more meaning to the participants and should be revealed to the reader of this report. The closing chapter focuses upon certain research thrusts which were best defined.

# ROLE OF EXPERIENCE IN DEVELOPMENT OF NEUROPHYSIOLOGICAL REGULATORY MECHANISMS AND IN ORGANIZATION OF THE BRAIN[1]

*Mark R. Rosenzweig*
University of California, Berkeley

This paper examines the concept that learning helps to organize and to reorganize the nervous system. The examination includes the following three main sections.

1. For a viewpoint rather different from that usually presented in psychology, the first section cites conclusions from the mathematical analysis of Bremermann to the effect that there is not enough genetic information to specify either bodily structure or behavior in full detail, so that some structural and behavioral organization must be acquired at the expense of a trial-and-error self-organizing period. Limitations of this viewpoint will be mentioned.

2. The next section considers the example of control of food intake in mammals. This reveals adequate organization from the time of birth but also shows changes in organization that permit the shift from dependent to independent behavior. The role of learning in the earliest phase will be stressed.

[1] The research reported in the third section of this paper was supported by NSF Grant GB-8011 and Office of Education Grant 09-140398. It also received support from the United States Atomic Energy Commission.

3. Measurable changes in the brain as consequences of differential experience will be described in the third section. Most of this work has been done in the postweaning period, but a recent preweaning study will also be presented. These cerebral changes may reflect processes of learning and memory storage; a number of alternative hypotheses have been ruled out. Therefore some of the brain changes may reflect the development of self-regulatory mechanisms.

The second and third sections will be based on research with infrahuman mammals. Following these sections there will be a brief discussion of the relevance of research with infrahuman animals to the study of human development.

## TRIAL-AND-ERROR SELF-ORGANIZATION AS A SUPPLEMENT TO LIMITED GENETIC INFORMATION

Bremermann (1967) has taken up mathematical aspects of genetics and of evolution considered as a goal-seeking self-organizing system. He reports that the mutation rate for fastest optimization of an evolving system is inversely proportional to the number of genes. In order to evolve speedily, a species should have as few genes as possible, and in this regard *E. coli* is superior to man. On the other hand, in order to be able to organize complex structures and behavior patterns, a species needs as many genes as possible. These two requirements are thus antagonistic, and different living systems represent various compromises. This implies that genes may be in short supply for organizing structure and behavior.

Bremermann attempts to estimate the amount of information that would be necessary to specify the details of neural nets in the brain, and he concludes that it would be greater than the genetic information available for the entire body. Since the brain cannot be specified in detail genetically, he suggests that it may either be organized according to rules and principles (that is, as a higher level global structure composed locally of repeating modules) or it could have a genetically controlled "master structure" with details left random and to be organized through interaction with the environment. I believe that examples of both types can be shown, but we will concentrate in this paper on interaction with the environment.

The informational "cost" of specifying items of behavior has also been estimated by Bremermann. He notes that releasers of behavior (in ethological terminology) are usually not highly detailed, and he concludes that "releaser stimuli are organized in such a way as to have as low a genetic cost as is consistent with their proper functioning." If elaborate innate movements were to be determined entirely genetically, the genetic cost would be high, but this cost is frequently avoided since "coordination in mammals and birds quite generally is not inborn, but acquired through self-organization."

"The *play instinct* of young mammals is delightful to watch in action. Its utilitarian purpose, however, is to provide a sequence of experiences that help to coordinate movements and organize behavior. Thus genetic information is saved at the expense of a trial-and-error self-organizing period."

*Possible Misleading Inferences*

It would be possible to draw misleading inferences from a formulation like that of Bremermann. Three that are related to our later discussion are pointed out here.

1. One such inference would be that brain and behavior are insufficiently organized at birth or in early days for the requirements to be made on them. Actually they must meet the needs at that age, or the young organism will not survive. We will take this up in the next section on the ontogeny of regulation of food intake. Anokhin (1964) has considered this problem at length in his discussions of "systemogenesis."

2. Another incorrect inference might be that brain and behavior become completely fixed at maturity. Not only can the adult continue to learn but the brain also continues to show plasticity, as we shall consider in the third section of this paper.

3. When animals have been kept in impoverished environments during early life, later inadequacies of structure or function have often been attributed to a lack of development because of insufficient early stimulation. In some cases, however, it now appears that organization was present earlier but degenerated during early deprivation of stimulation. Thus Hubel (1967) has demonstrated that much detailed organization is present at birth in the visual system of mammals, but that this decays if visual experience is prevented in early life. Similarly, Lessac and Solomon (1969) have shown that behavior capacities present at 12 weeks of age in puppies become disorganized during subsequent isolation. Such examples of initial genetic specification plus the need for stimulation would not have been predicted from Bremermann's schema.

## ONTOGENY OF REGULATION OF INTAKE OF FOOD

### Description of Early Development and Behavior

The ontogeny of physiological and behavioral mechanisms controlling the intake of food and water has been studied in several infrahuman mammals. I will draw material principally from studies of the rat but will supplement this at certain points with findings made on other species. The ontogeny of the physiological mechanisms has been reviewed recently by Kennedy (1967).

From birth to about day 14 the rat pup ingests only milk and refuses water or solid food. Thereafter it can eat solid food and drink water, although normally weaning is complete only at about 30 days of age. At 14 days the rat, which is born poikilothermic, has developed sufficient regulation of temperature to be able to leave the nest and begin to forage. It is just at this stage that the rat can begin to use its distance receptors effectively, as Gottlieb (1971) has reviewed recently. The ear canal opens about day 12-13 and the eyelids open about day 14-15. Some responses to strong auditory and visual stimuli can be elicited a few days earlier when the cochlea and retina become relatively mature. Adult regulators of fluid and electrolytes also begin to appear at about day 14. The kidneys of most mammals are immature structurally and functionally at birth, but feeding milk does not embarrass the kidney, since at this age most of the nitrogen is taken up by cellular growth. Normal plasma levels of sodium, potassium, and phosphate are also maintained without the complex regulatory devices of the adult. This is probably accomplished "by self-regulation at the cellular level, so long as a proper substrate for growth is provided."

The rate of growth of rat pups is extremely rapid. In litters of 3, they gained about 200% in body weight the first week; in litters of 15, about 80%. During the first week, intake per gram of body weight is only about as great as that of the adult, but weight gain is rapid because the pup stays in the warm nest and expends almost no calories for temperature regulation. In the next weeks the young rats eat proportionately more but grow less rapidly; their intake per unit of body weight is greater than that of adults. The food intake of the mother rats also increases to two or three times the normal level and equals the highest intakes recorded after hypothalamic lesions in nonlactating rats. Since the rat pups increase their intake and weight almost exponentially, the mother cannot continue to match this rate of increase by eating still more. Shortage of milk must be an important factor in causing natural weaning of the pups, and they gradually change to dry food as the supply of milk falls progressively short of their demands.

Early in the period of lactation the female retrieves pups that leave the nest, but this usually stops by the twentieth day after birth (Wiesner & Sheard, 1933). Retrieval depends upon the age of the pups. Offered pups of different ages on separate trials, a mother retrieves pups up to a certain age, with different females showing different thresholds of age. It used to be supposed that the female was discriminating on the basis of vision or of a sensory pattern involving touch and smell as well as sight. Now it has been found that in mice, which show similar behavior, retrieval is stimulated by the ultrasonic cries of the young (Noirot & Pye, 1969). From about day 4 on, there is a regular decrease in the length of these calls, their bandwidth, and their sound pressure.

One of the few recent studies of ontogeny of behavior in the laboratory rat, undisturbed by any experimental manipulation, is that of Bolles and Woods

(1964). They observed that the first interest in hard food begins with sniffing it on day 15. By day 16 the pup may sit on its haunches in the adult eating posture nibbling a food pellet held in its forepaws. On successive days, more and more time is spent eating, and the eating is persistent; a young rat may nibble on a food pellet for as long as 10 min. A few days after weaning (which occurred artificially at 21 days in this study), young rats may try to take food from each other, even when it is abundant on the cage floor. Soon, thereafter, a rat tries to get away from other rats when it is eating. Since several often eat at once (this is social facilitation) the litter tends to become distributed more or less evenly around the cage.

An unplanned observation of Bolles and Woods was perhaps the first indication of learning in rats during the first days of life; we will soon point out fuller and more conclusive observations made by Thoman, Wetzel, and Levine (1968). Bolles and Woods noted the case of one litter whose mother characteristically stood over the young while nursing instead of adopting the usual crouching posture. On day 5 when the mother was away from the nest, several of her pups "were observed to raise their heads and wave them around in searching-like behaviour. Such behaviour was not observed in other litters" (p. 441). Bolles and Woods stated that if this was not just a chance occurrence, then it may be "an interesting finding in view of the difficulty usually reported in obtaining conditioning or other learning in infant rats."

*Effects of Early Weaning*

Křeček (1971) used radioactive tracers to test when rats first start eating food that is available in the cage and when they last take any milk from the mother. His animals began to eat solid food at about 14 days of age but did not completely stop nursing until about day 28. Křeček points out that by this standard most laboratory rats are weaned prematurely. Moreover, Nováková (1966) has reported that premature weaning has deleterious effects on later formation of conditioned responses. Even rats weaned at day 21 were found to be somewhat inferior to those separated from the mother at day 28. Early weaning also produces measurable effects on RNA content of brain cells (Nováková *et al.*, 1969). The effects of early weaning on later conditioning can be prevented if the pups are given both a high fat diet and the company of a nonlactating adult female; the high fat diet alone will make up only part of the deficit.

(Denenberg reported at this Conference that the effects of early weaning had been debated at the Conference on Postnatal Development of Phenotype held in Liblice, Czechoslovakia, in 1967. He stated that Levine, in unpublished studies, had not found significant differences on behavioral tests between rats weaned at 21 or 30 days.)

## Role of Learning in Early Feeding

Most discussions of the ontogeny of regulation of food intake omit any reference to the role of learning, even though the importance of learning for later establishment of food preferences is acknowledged. As we shall see, puppies and kittens need to learn in their first days in order to get to the nipple and suckle effectively; this is probably true also for the rat, although it does not seem to have been tested yet.

How heredity and environment contribute to sucking (both nursing and nonnutritive sucking) is an old question whose history has been reviewed in an interesting chapter by McKee and Honzik (1962), "The sucking behavior of mammals: An illustration of the nature-nurture question." They cite William Harvey as writing in 1651 that Hippocrates had held that the fetus must have practiced sucking in the womb or else it would not be able to suck so efficiently as soon as it is born. Harvey supported this with the further observation that the efficiency of sucking by an infant decreases when sucking is discontinued for only a few days.

At what age learning first appears in mammals depends upon what learning situation is used, as Rosenblatt (1971) has reviewed. Conditioned withdrawal to shock or to tactile stimuli cannot be produced in the dog before the second or third week, and such results have been used to locate the critical period of socialization in the dog. On the contrary, in the area of suckling and related activities which must be present or develop very early, learning can be demonstrated in the first days of a puppy's life. Thus Stanley *et al.* (1963) showed conditioning of either sucking or struggling beginning on day 3, depending upon whether the puppy had previously been given milk or quinine solution after insertion of a clean nipple into the mouth. Russian investigators have reported conditioned sucking responses as early as day 1 in puppies and rabbits. Rosenblatt has reported detailed studies from his laboratory in which newborn kittens learn to feed independently in 2 or 3 days from a special brooder. They also learn to discriminate by the tactual forms of the flanges at the bases of the rubber nipples. The textures of the flanges are varied, with concentric circles or raised dots. A kitten soon sucks from the one that has provided it milk and rejects the other. Olfactory discrimination can similarly be shown by day 3. When nursing from their mother, kittens form preferences for a particular pair of nipples during the first few days. If a kitten is on its preferred nipple, it will remain there when another kitten approaches; if it is on a nonpreferred nipple, it will yield when another kitten comes and tries to take the nipple. This behavior is shown by 3-4 day-old kittens. The mother cat confines the young kittens between her legs and her body, thus facilitating their finding the nipples. The kittens must learn to reach and grasp the nipples, and the effective working of the nursing pattern requires rather complex learning on the part of the kitten.

Incubator rearing of rats has recently been demonstrated to be feasible by Thoman and Arnold (1968a). The pups were kept, either singly or in a group, in a warm humid incubator which was provided with a warm moist pulsating tube as a "mother surrogate." The pups were fed with a special stomach catheter. After about two days of this feeding, the subjects facilitated the insertion of the catheter "by searching for, licking, and swallowing the tube immediately" (p. 442). In a follow-up study (Thoman, Wetzel, & Levine, 1968) it was found that incubator-reared rats at 3 days of age did not suckle or even root on an immobilized lactating female rat, whereas in response to stimulus components of the artificial feeding situation they showed a striking increase in activity level, along with suckling mouth movements and squealing. Mother-reared pups of this age were generally unresponsive to the artificial feeding stimuli. Thus variations in rearing conditions apparently provided opportunities for the acquisition of different behavior patterns during the first few postnatal days.

A few further results of Thoman and Arnold deserve at least brief mention here. When the incubator-raised and normally raised female rats became young adults, they were impregnated and their subsequent maternal behavior was observed and rated. There were no significant differences between groups in size and weight of litters, or in ratings of nest-building or retrieval. But in spite of the fact that observation revealed no differences in maternal behavior, the normally bred rats proved to be more effective mothers. Fewer of their offspring died, and more had their eyes open at 15 days than did the pups of incubator-raised mothers (1968a). If age of opening is a reflection of the amount of stimulation provided to the pup, then normally reared mothers gave their pups more stimulation than did the artificially reared animals. In another paper, Thoman and Arnold (1968b) measured behavior of incubator-reared and normally reared rats in an open field test and in problem-solving behavior in the Hebb-Williams test. No significant differences were found (perhaps in part because only 8 animals of each type were used), but the incubator animals averaged somewhat better on Hebb-Williams scores. At least, "the incubator procedures did produce animals without behavioral deficiency" (1968b, p. 222).

### *Long CS–US Delays with Taste as CS*

There has been a problem in explaining the role of learning in the highly accurate regulation of intake of food and water, as Rozin (1969) has discussed. The problem is that there is an inherently long delay between the ingestion of a substance (CS) and the fundamental metabolic effects produced by the substance (US), whereas most data and theories concerning animal learning emphasize the critical importance of close temporal contiguity between CS and US. Nevertheless, learning mechanisms bridging relatively long intervals have been implicated and assumed to be present in caloric regulation (Le Magnen, 1967), in specific hungers (Harris, Clay, Hargreaves, & Ward, 1933; Rozin, 1967) and in food aversions based on X-radiation or poisoning (Garcia, Kimeldorf, & Hunt, 1961; Rzóska, 1953). It has been shown that aversions with long CS–US intervals can be established much more easily with taste as the CS than with other cues, and it has been suggested that aftertastes may provide peripheral mediation of the long-delay interval. Rozin (1969) has demonstrated that rats can learn to avoid either the higher or the lower concentration of the same substance; they avoid the concentration that is followed after 30 min by injection of a drug that produces unpleasant consequences, probably including nausea. The strength of the aversion increases over trials on successive days. Since the aftertastes of the two concentrations are unlikely to differ 30 min later, the results suggest that the long-delay learning is mediated centrally.

#### *Diet and Nonspecific Excitatory Level*

Lát (1967, 1969) has reported reciprocal relations between diet and what he calls "Nonspecific Excitatory Level" (NEL) in the rat. NEL is measured by a weighted average of various activities, with the greatest weights given to rearing up on the hind legs and to locomotion. This index correlates well with hippocampal theta activity. NEL also shows different relations with various behavioral measures.

Rats that differ in NEL also tend to differ in dietary preferences, high NEL rats preferring a "glucogenic" diet (one high in carbohydrate), whereas low NEL rats show a "gluconeogenic" intake pattern (one relatively high in protein and fat but low in carbohydrate) (1967). Conversely, it is possible to alter NEL by giving different diets. High-carbohydrate, high-fat, and high-salt diets were reported to increase NEL, whereas a high-protein and high-potassium diet decreased NEL. When two groups were given either high-carbohydrate or high-protein diets from 12 to 52 days of age, differences in NEL were still present at 200 days of age (1969). Lát concludes, "In the nature and nurture dichotomy, the term 'nurture' is justified in both meanings: in the nutritional, as well as in the educational, one" (1969, p. 719).

### Maturation of Brain Mechanisms Controlling Intake

Until around the time of puberty, the brain mechanisms controlling intake may not function in a completely adult fashion in the rat, as Wade and Zucker (1970) have demonstrated. Normally estrogen depresses food intake and body weight in adult female rats, and these effects can be obtained with doses smaller than those that are necessary to affect sex behavior. The effects on sex behavior can, however, be obtained at an earlier age than the influences on intake, indicating dissociation of the mechanisms involved. When rats were ovariectomized or sham operated on day 1, the weights of the two groups remained nearly identical through day 42. On the next weighing at day 49, the spayed rats were 10% heavier than the controls, and the weights of the two groups continued to diverge from then on. The fact that there was no difference earlier was not due simply to lack of estrogen, because neither sexual precocity nor exogenous estrogen could induce a drop in food intake or in weight gain before about day 40. Rather, it appears that growth hormone from the pituitary induces a "hyperphagia" that cannot be overcome by satiety signals from the ventromedial hypothalamus (VMH). Growth hormone may also make VMH less responsive to estrogen. If the pituitary is removed in prepubertal females, then administration of estrogen does decrease food intake. In this case it appears that the neural mechanism is ready earlier but can come into play only when the secretion of growth hormone declines at about 40-50 days of age.

The ontogenetic development of feeding behavior in the rat has been reported to parallel in detail the stages of recovery observed after damage to the lateral hypothalamus (LH) in the adult, according to Teitelbaum, Cheng, & Rozin (1969). How much this research tells us about the ontogenesis of feeding merits our scrutiny. The investigators slowed down rapid early development by thyroidectomizing rats on the first or second day after birth and then studied their behavior at weaning at 21 days of age. Each of the 35 animals could be classified into one of the four stages that have been described for recovery from LH damage. (The stages, in brief, are the following: I. Aphagia and adipsia; II. Anorexia and adipsia; III. Adipsia; IV. Relative recovery but abnormalities still shown on sensitive tests of regulation.) The 12 animals that subsequently changed from one stage to another always progressed to the next higher stage of recovery. Three animals progressed from Stage IV to full recovery. Teitelbaum *et al.* conclude, "Because the lateral hypothalamic syndrome can be reinstated after recovery by spreading cortical depression, and because recovery of motor function in humans and monkeys can also be seen to parallel development, we suggest that lateral hypothalamic recovery is essentially a process of reencephalization of function. This could explain why recovery recapitulates ontogeny" (p. 440). At the time of the Conference I raised the following question with regard to this argument: If ontogeny of control of feeding involves the organization of a neural system in which LH plays an important role, why should reorganization after destruction of LH reproduce in detail the organization of the original system?

Shortly after the Conference, an important article by Powley and Keesey (1970) brought into question the conclusions of Teitelbaum about stages of recovery after damage to LH. They found that if the body weight of young adult rats was reduced about one sixth by underfeeding prior to lesioning the hypothalamus, then "recovery" usually occurred in less than 1 day. Furthermore, many operated rats actually showed hyperphagia until they reached a new weight plateau. This postoperative level was lower than their previous normal weight but higher than the underfed weight. Powley and Keesey suggest that an animal's set-point for weight is determined by a balance between activity of the lateral (excitatory) and ventromedial (inhibitory) hypothalamic nuclei. After lesioning, an animal will attempt to reach the new set-point by either feeding or refusing to eat, depending upon whether its body weight is below or above the target value. In other words, the self-regulatory mechanism still functions after lesioning, but with the balance tipped in one direction or the other depending upon the site of the lesion. In the case of the study by Teitelbaum *et al.* with early thyroidectomy, this procedure also determines slower growth and a reduced body weight. The impaired eating behavior seen in these animals may then also represent response to an altered set-point of the regulatory mechanism.

## MEASURABLE CHANGES IN BRAIN AS CONSEQUENCES OF DIFFERENTIAL EXPERIENCE

### Effects in the Postweaning Period

Differential experience produces measurable changes in brain weights, brain chemistry, and brain anatomy, as our group in Berkeley has shown in a number of reports since 1960. This project represents a longstanding collaborative effort headed by Dr. Edward L. Bennett, a biochemist, Dr. Marian C. Diamond, a neuroanatomist, and myself, a physiological psychologist. Several other investigators have also found changes in brain chemistry and anatomy as consequences of differential experience, as we have reviewed recently (Rosenzweig & Leiman, 1968; Rosenzweig, 1971; Bennett & Rosenzweig, 1971). Let me summarize briefly some of our findings.

As compared with littermates kept in standard colony conditions (SC), rats put at weaning into an environmentally enriched condition (EC) for 30 days or more show increased weight of cerebral cortex, especially in the occipital region, increased thickness of cortex, and increased total activities of acetylcholinesterase, cholinesterase, and hexokinase. EC consists of placing rats in a group of 10-12 in a large cage with stimulus objects that are changed daily. Rats put into an impoverished condition (IC) show changes in the opposite direction from the SC baseline. IC consists of housing rats in isolation or, in a few experiments, two in a small cage. EC–IC differences found in the occipital region are summarized in Table 1. This table is based on results with male $S_1$ strain rats exposed to EC or IC starting at 25 days of age and sacrificed at 105 days. In the top row of Table 1 we see that in the weight of a standard sample of occipital cortex, the EC mean exceeded the IC mean by 6.2% ($p < .001$). This result is based on 14 experiments which included 151 littermate pairs. In 114 of the 151 pairs–three quarters of the cases–it was the EC rat that had the greater weight of occipital cortex. In a separate experiment we have demonstrated that the percentage weight difference obtained with fresh tissue is also found after the tissue has been dried, so the weight differences that we have reported previously cannot be attributed to changes in fluid content (Bennett, Rosenzweig, & Diamond, 1969).

TABLE 1

*Effects of Differential Experience on Occipital Cortex of Rats Kept in Enriched or Impoverished Conditions from 25 to 105 Days of Age*

| Measure | % Difference | $p$ | No. EC > IC |
|---|---|---|---|
| Weight | 6.2 | < .001 | 114/151 |
| Total protein | 7.8* | < .001 | 25/32 |
| Thickness | 6.3 | < .001 | 47/52 |
| Total AChE | 2.6 | < .01 | 90/150 |
| Total ChE | 10.2 | < .001 | 86.5/108 |
| Total hexokinase | 6.9** | < .01 | 17/21 |
| Number of neurons | −3.1 | NS | 7/17 |
| Number of glia | 14.0 | < .01 | 12/17 |
| Perikaryon cross section | 13.4 | < .001 | 11.5/13 |

*Weight difference 7.0% in these experiments.
**Weight difference 5.5% in these experiments.

It should be noted that the greater weight of cerebral cortex in the EC rats does not reflect greater body weight. On the contrary, at the end of the 25- to 105-day period, the $S_1$ EC rats weigh about 10% less than do their IC littermates.

Percentage differences between littermate EC and IC rats on a number of chemical and anatomical measures are also presented in Table 1. In all cases the rats were sacrificed under code numbers that did not reveal whether they were EC or IC to the people who weighed the brain samples and who performed the chemical and anatomical analyses. The details of analytical methods have been presented in the original reports (Rosenzweig, Krech, Bennett, & Diamond, 1962; Rosenzweig, Bennett, Diamond, Wu, Slagle, & Saffran, 1969; Diamond, 1967). For some of the measures reported in Table 1 the numbers of subjects are rather small, and replications of some of these experiments are now being run. The table demonstrates that the brain can be altered in many regards by differential environmental treatment, and further dimensions of change will doubtless be found as still other types of measures are taken.

Experiments which have not yet been completely analyzed indicate that the number of dendritic spines in pyramidal cells in the occipital cortex is greater in EC than in IC rats. This work is being done in collaboration with Dr. Albert Globus. It is not so much that the spines are more numerous per unit of length of dendrite but chiefly that, with increased cortical thickness, the dendrites are longer and carry a greater total number of spines.

It must be conceded that neither we nor other investigators have clearly demonstrated altered neural circuits as a consequence of differential experience. The increased number of dendritic spines points in this direction, as do reports of changes in synaptic number and dimensions with altered visual experience, both in the retina (Cragg, 1969a) and in the thalamus (1969b). More work is obviously needed to test the hypothesis that the cerebral changes with experience represent, in part at least, processes of learning and memory storage.

It is easier to produce these effects than we had originally supposed. As Table 2 shows, reducing the exposure period from 80 to 30 days does not diminish the weight effects. In the occipital region, the difference is actually larger for the shorter period. Further reducing the treatment to only 2 hours per day for 30 days still leaves the EC–IC differences intact. The highly stable measure of the ratio of cortical weight to that of the rest of the brain yields substantially equal results for all three durations of differential experience.

Similar changes can be produced in rats given differential experience only in adulthood (Rosenzweig, Bennett, & Krech, 1964; Riege, in press). Thus the brain retains plasticity, and effects of experience are not confined to early life.

A series of experiments reported elsewhere (Rosenzweig, 1966; 1970; 1971) has ruled out a number of alternatives to differential experience as causes of these cerebral effects. Thus, we demonstrated in 1960 (Krech, Rosenzweig, & Bennett) that these effects cannot be attributed to differential handling, so that they differ from the effects to be described in the next paper by Denenberg and Zarrow. Other control experiments have shown that our cerebral effects are not caused by locomotor activity, accelerated maturation, differential endocrine activity or by stress. Visual stimulation is not required to produce these effects even in the occipital cortex, but the occipital area is known in the rat to receive inputs not only from vision but also from several other sensory modalities. Neither social stimulation alone nor stimulus objects alone produced significant effects, but the combination of social and inanimate stimuli did alter brain values. Further research is underway to separate out effects of these sources of experience.

It is possible that feeding and drinking enter into the production of these brain effects. A series of experiments performed at Cornell (e.g., Gibson & Walk, 1956; Gibson, Walk, & Tighe, 1959) has demonstrated that if rats are exposed to visual forms in their cages over a period of weeks—for example, to a triangle or to a circle—they can later learn more readily to discriminate these forms than can nonexposed control animals. Recently Kerpelman (1965) and Bennett and Ellis (1968) have confirmed these results but have shown that they may occur only if the experience of the visual form is reinforced; that is, only if food and water were present during the exposure did the effect appear clearly. Further work is now being done to determine whether reinforcement is a strict requirement for this sort of learning.

TABLE 2

*Comparison of Percentage Differences between EC and IC Brain Weights from Experiments of Various Durations*

| | | | |
|---|---|---|---|
| Duration (days) | 80 | 30 | 30 |
| Hours per day in EC | 24 | 24 | 2 |
| No. experiments | 14 | 13 | 3 |
| No. pairs | 151 | 135 | 28 |
| Occip. cortex | 6.2** | 10.4** | 15.0** |
| Total cortex | 4.1** | 5.9** | 4.1** |
| Rest of brain | −1.0* | 0.5 | −0.4 |
| Cortex/rest | 5.2** | 5.3** | 4.6** |

Note.—Male $S_1$ rats; experiments started at about 25 days of age.
*$p < .01$.
**$p < .001$.

The basic EC—IC cerebral effects have been found not only in the several strains of rats that we have tested but also in laboratory mice (La Torre, 1968) and in gerbils (Rosenzweig & Bennett, 1969). Thus these brain effects seem to occur rather generally, at least among laboratory rodents.

## Cerebral Effects in the Preweaning Period

We have hesitated to work with the rat before the age of weaning because, as noted above, it is quite fragile. Until the age of about 14 days, the rat pup does not have adequate temperature regulation and it cannot use its eyes and ears for distance perception. It is hard enough to extrapolate from infrahuman animals to man, but, as Eayrs (1968) has stressed, the rat in its first week is roughly at the stage of the human fetus in the fifth month of gestation. Therefore the effects of manipulations undertaken on the rat during its first week could be related only to intrauterine manipulations of the human fetus. This does not mean that such early manipulations lack interest, and Eayrs himself has studied effects of thyroidectomy during the first days of life on development of the rat brain. It does raise the question whether effects of treating the rat during its first postnatal days should be considered in terms of sensory stimulation and experience or rather, as Denenberg and Zarrow do,[2] in terms of direct alteration of neuroendocrine mechanisms.

Malkasian, in his doctoral research (1969), has recently extended some of our anatomical measures to the preweaning period. He took litters of Long–Evans rats, reduced each litter to 3 male pups, and put the litters into one of

[2] See chapter by Denenberg and Zarrow, this volume page 39.

three conditions when the pups were 6 days of age: (a) unifamily environment (UFE), that is, one mother and her pups in a colony cage; (b) multifamily environment (MFE), three litters in a large empty cage; (c) enriched condition multifamily environment (EC–MFE), three litters in a large cage with stimulus objects that were changed daily.

The main results for measurement of cortical thickness appear in Table 3. EC–MFE showed significant differences from UFE, so either the stimulus objects or the combination of social and inanimate stimuli altered the brain values. At 14 days, when the eyes were just opening, occipital cortex did not show as clear an effect as did somatosensory cortex. By 19 days, occipital cortex showed somewhat larger effects than somatosensory cortex, and this was also true at 28 days. EC–MFE also exceeded UFE at 28 days in cross-sectional area of neuronal nuclei in the occipital and somatosensory regions. Note, in the bottom line on the table, that at 28 days of age MFE did not differ significantly from UFE, so added social stimulation was not by itself capable of altering cortical depth.

A further experiment was run in which the EC–MFE group was given daily injection of reserpine to determine whether this tranquilizer would prevent occurrence of the increased thickness of cortex. These animals were compared with UFE animals that were run concurrently. Results for somatosensory cortex are presented in Table 3 (EC–MFE–Reserp. minus UFE). The drug did not prevent differences from appearing in the medial and intermediate aspects of the cortex, although there was no effect in the lateral region. Measures were not made in occipital cortex, because injections of reserpine in young rat pups was found to irritate the conjunctiva and to delay opening of the eyes.

The overall results of Malkasian's study demonstrate that enriching the environment from day 6 on produces measurable effects on brain anatomy. This is achieved without handling the pups or stressing them. The results indicate that brain organization is affected by experiential variables from the time that the rat first begins to leave the nest and expose itself to the external environment.

The rapid early changes in brain values of animals under all the experimental conditions may also reflect changes in neural systems necessary to the maintenance of life. New investigations will be needed to test this hypothesis.

## DISCUSSION

The relevance of animal research for human studies deserves explicit consideration at this conference of the Institute for Study of Human Behavior, especially since two papers presented here are based upon animal experiments. Although I discussed this question before the Division of Developmental Psychology a few years ago (1966), I would like to add some further considerations in the context of this conference.

TABLE 3

*Percentage Differences in Cortical Thickness Measures between Rats Assigned to Differential Environments at 6 Days of Age*

| | Ns | Somatosensory Cortex | | | Occipital Cortex | | |
|---|---|---|---|---|---|---|---|
| | | Medial | Intermed. | Lateral | Medial | Intermed. | Lateral |
| EC-Multifamily envir. minus unifamily envir. | | | | | | | |
| 14 days | 7,7 | 10.6 *** | 10.0 *** | 7.0 * | 1.5 | 6.8 | 16.0 ** |
| 19 | 6,9 | 5.5 *** | 9.7 *** | 7.6 | 6.8 * | 8.0 * | 14.1 ** |
| 28 | 19,19 | 8.0 ** | 6.9 ** | 9.0 *** | 9.1 ** | 10.3 *** | 11.9 *** |
| EC-MFE-Reserp. minus UFE | | | | | | | |
| 28 days | 12,12 | 7.0 *** | 5.0 ** | −1.5 | – | – | – |
| MFE minus UFE | | | | | | | |
| 28 days | 12,12 | 0.3 | 2.8 | 0.9 | 3.6 | 0.9 | 1.7 |

Note.–From Table 14a, p. 74, Malkasian thesis, with some changes and additions.
*$p < .05$.
**$p < .01$.
***$p < .001$.

As one who has experimented chiefly with infrahuman subjects, I do not want to claim that the results of such research can be applied directly to human beings. Even extrapolations of animal findings to people must be made cautiously and can be considered only tentative, to be confirmed or rejected by observations or experiments made on human subjects. Turning now to the positive side, not only can such extrapolations serve as valuable guides to human research, but many of the concepts and techniques that have originated from animal research can also be adapted profitably to human studies.

A few illustrations will serve to show that findings made with one animal species are often difficult to apply to closely related species, let alone to distant ones. Thus, the effects of isolation on emotional behavior are quite different in the mouse and in the rat. The effects on dogs differ from the rodent patterns, and different strains of dogs display their own characteristic types of responses. Isolation produces still different effects on the emotional behavior of the rhesus monkey. In the case of intellectual ability, too, I have presented elsewhere a rather detailed discussion to indicate that we cannot–at least at present–draw any clear generalization about effects of environmental enrichment or impoverishment from studies of a number of animal species (1971).

If results of animal experiments can rarely be applied directly to man, many examples show that concepts and techniques from animal research can enrich the study of human behavior. Earlier in this paper we discussed the contributions of learning to the development of self-regulatory mechanisms. It is noteworthy that many of the investigators who have studied human learning have themselves done research in animal learning or have derived concepts and methods from animal experimentation. Thus when Thorndike published his striking thesis on animal intelligence in 1898, stressing the role of "trial and error and accidental success," Cattell urged him to apply his methods to the study of children and young people. And Thorndike began in this way a productive career of four decades at Teachers College, Columbia. A decade later Köhler made his observations on insightful problem-solving in chimpanzees, and this led to Gestalt studies of productive thinking in human subjects. At about the same time, John B. Watson used the animal studies of Pavlov (as well as the human experiments of Bekhterev) to demonstrate the potentialities of objective studies of behavior. Watson, himself an investigator of animal learning, then fostered the early study of conditioning in children. As well as classical conditioning, instrumental (type II) conditioning (Konorski), or operant conditioning (Skinner), was first studied in animals and has continued to be used extensively with infrahuman subjects. The basic techniques worked out in these studies have recently been employed in experiments on development of learning in very young children. It seems clear that many of the concepts and techniques of studying learning that had their origin in animal research have since enriched the study of human behavior.

It must be acknowledged that there has been, and continues to be, resistance against the application of insights or techniques from animal research to human studies. Some appear to feel that it may debase people to apply to them techniques that have been derived from the animal laboratory. Part of the criticism of Watson and of Skinner has stemmed from this source.

In other fields of development, too, there seems to be much that animal studies can suggest for observations and experiments on human subjects. Even if neither the altricial rat nor the precocial guinea pig offers a model for human development, factors that are found to be important at certain developmental stages (or at all stages) in these and other species may suggest fruitful lines of investigation in man.

## CONCLUSIONS

The information needed for the elaboration of self-regulatory mechanisms is partly genetic and partly in the environment. The ontogeny of regulation of food intake was considered as an example of development of self-regulatory mechanisms. At each stage the behavior and physiology of the young mammal are appropriately organized to meet the demands made upon them. Learning is necessary for organization of some aspects of suckling, and learning in the feeding situation can be demonstrated during the first days after birth; this is earlier than had been found with other types of response. The brain of the rodent has been demonstrated to respond to differential environmental treatment by changes in tissue weight, biochemistry, and anatomy. Such changes occur in adult as well as in immature animals. The hypotheses were proposed that some of the cerebral changes that occur as consequences of differential experience reflect the development of self-regulatory mechanisms and that some may also reflect other processes of learning and memory storage. Testing these hypotheses will require much further research.

## REFERENCES

Anokhin, P. K. Systemogenesis as a general regulator of brain. In W. A. Himwich & H. E. Himwich (Eds.), *The developing brain.* Vol. 4. *Progress in brain research.* Amsterdam: Elsevier, 1964.

Bennett, E. L., & Rosenzweig, M. R. Chemical alterations produced in brain by environment and training. In A. Lajtha (Ed.), *Handbook of neurochemistry,* Vol. 6. New York: Plenum Press, 1971.

Bennett, E. L., Rosenzweig, M. R., & Diamond, M. C. Rat brain: effects of environmental enrichment on wet and dry weights. *Science,* 1969, vol. 163, 825-826.

Bennett, T. L., & Ellis, H. C. Tactual-kinesthetic feedback from manipulation of visual forms and nondifferential reinforcement in transfer of perceptual learning. *Journal of Experimental Psychology,* 1968, vol. 77, 495-500.

Bolles, R. C., & Woods, J. B. The ontogeny of behaviour in the albino rat. *Animal Behaviour,* 1964, vol. 12, 427-441.

Bremermann, H. Quantitative aspects of goal-seeking self-organizing systems. In F. M. Snell (Ed.), *Progress in theoretical biology,* Vol. 1. New York: Academic Press, 1967.

Cragg, B. G. Structural changes in naive retinal synapses detectable within minutes of first exposure to daylight. *Brain Research,* 1969, vol. 15, 79-96. (a).

Cragg, B. G. The effects of vision and dark-rearing on the size and density of synapses in the lateral geniculate nucleus measured by electron microscopy. *Brain Research,* 1969, vol. 13, 53-67. (b).

Denenberg, V. H., & Zarrow, M. X. Effects of handling in infancy upon adult behavior and adrenocortical activity: suggestions for a neuroendocrine mechanism. This volume, page 39.

Diamond, M. C. Extensive cortical depth measurements and neuron size increases in the cortex of environmentally enriched rats. *Journal of Comparative Neurology,* 1967, vol. 131, 357-364.

Eayrs, J. T. Developmental relationships between brain and thyroid. In R. P. Michael (Ed.), *Endocrinology and human behaviour.* London: Oxford University Press, 1968.

Garcia, J., Kimeldorf, D. J., & Hunt, E. L. The use of ionizing radiation as a motivating stimulus. *Psychology Review,* 1961, vol. 68, 383-395.

Gibson, E. J., & Walk, R. D. The effect of prolonged exposure to visually presented patterns on learning to discriminate them. *Journal of Comparative and Physiological Psychology,* 1956, vol. 49, 239-242.

Gibson, E. J., Walk, R. D., & Tighe, T. J. Enhancement and deprivation of visual stimulation during rearing as factors in visual discrimination learning. *Journal of Comparative and Physiological Psychology,* 1959, vol. 52, 74-81.

Gottlieb, G. Ontogenesis of sensory function in birds and mammals. In E. Tobach (Ed.), *Biopsychology of development.* New York: Academic Press, 1971.

Harris, L. J., Clay, J., Hargreaves, F., & Ward, A. Appetite and choice of diet: The ability of the vitamin B deficient rat to discriminate between diets containing and lacking the vitamin. *Proceedings of the Royal Society, London, Series B,* 1933, vol. 113, 161-190.

Hubel, D. H. Effects of distortion of sensory input on the visual system of kittens. *The Physiologist,* 1967, vol. 10, 17-45.

Kennedy, G. C. Ontogeny of mechanisms controlling food and water intake. In C. F. Code (Ed.), *Handbook of physiology,* Sec. 6, Vol. I. *Alimentary canal.* Washington, D.C.: American Physiology Society, 1967.

Kerpelman, L. C. Preexposure to visually presented forms and nondifferential reinforcement in perceptual learning. *Journal of Experimental Psychology,* 1965, vol. 69, 257-262.

Křeček, J. The theory of critical developmental periods and postnatal development of endocrine functions. In E. Tobach (Ed.), *Conference on biopsychology of development.* American Museum of Natural History, 1971.

Krech, D., Rosenzweig, M. R., & Bennett, E. L. Effects of environmental complexity and training on brain chemistry. *Journal of Comparative and Physiological Psychology,* 1960, vol. 53, 509-519.

Lát, J. Self-selection of dietary components. In C. F. Code (Ed.), *Handbook of physiology,* Sec. 6, Vol. 1. *Alimentary canal.* Washington, D.C.: American Physiological Society, 1967.

Lát, J., & Gollová-Hémon, E. Permanent effects of nutritional and endocrinological intervention in early ontogeny on the level of nonspecific excitability and on lability (emotionality). In E. Tobach, P. D. Albertson, and M. Krauss (Eds.), *Experimental approaches to the study of emotional behavior. Annals N.Y. Acad. Sci.,* 1969, vol. 159, 710-720.

La Torre, J. C. Effect of differential environmental enrichment on brain weight and on acetylcholinesterase and cholinesterase activities in mice. *Experimental Neurology,* 1968, vol. 22, 493-503.

Le Magnen, J. Habits and food intake. In C. F. Code (Ed.), *Handbook of physiology,* Sec. 6, Vol. 1, *Alimentary canal.* Washington, D.C.: American Physiological Society, 1967.

Lessac, M. S., & Solomon, R. L. Effects of early isolation on the later adaptive behavior of Beagles: a methodological demonstration. *Developmental Psychology,* 1969, vol. 1, 14-25.

Malkasian, D. R. The morphological effects of environmental manipulation and litter size on the neonate rat brain. Unpublished doctoral thesis, University of California, Berkeley, 1969.

McKee, J. P., & Honzik, M. P. The sucking behavior of mammals: An illustration of the nature-nurture question. In L. Postman (Ed.), *Psychology in the making.* New York: A. A. Knopf, 1962.

Noirot, E., & Pye, D. Sound analysis of ultrasonic distress calls of mouse pups as a function of their age. *Animal Behavior,* 1969, vol. 17, 340-349.

Nováková, V. Weaning of young rats: Effect of time on behavior. *Science,* 1966, vol. 151, 475-476.

Nováková, V., Schlutter, G., & Sandritter, W. RNS-Gesamtmenge der Ganglienzellen des Zentralnervensystems bei normal und frühentwöhnten Ratten. *Virchows Archiv, Abteilung B, Zellpathologie,* 1969, vol. 4, 16-20.

Powley, T. L., & Keesey, R. E. Relationship of body weight to the lateral hypothalamic feeding syndrome. *Journal of Comparative and Physiological Psychology,* 1970, vol. 70, 25-36.

Riege, W. H. Environmental influences on brain and behavior of old rats. *Developmental Psychobiology* (in press).

Rosenblatt, J. S. Early ontogeny of modification with special reference to the development of suckling in newly born kittens. In E. Tobach (Ed.), *Conference on biopsychology of development.* American Museum of Natural History, 1971.

Rosenzweig, M. R. Evidence for anatomical and chemical changes in the brain during primary learning. Symposium 20, International Congress of Psychology, Moscow, 1966. Pp. 5-17. [Revised version in K. Pribram and D. Broadbent (Eds.), *Biology of memory.* New York: Academic Press, 1970, 69-85.]

Rosenzweig, M. R. Effects of environment on development of brain and of behavior. In E. Tobach (Ed.), *Biopsychology of development.* New York: Academic Press, 1971.

Rosenzweig, M. R., & Bennett, E. L. Effects of differential environments on brain weights and enzyme activities in gerbils, rats, and mice. *Developmental Psychobiology,* 1969, vol. 2, 87-95.

Rosenzweig, M. R., Bennett, E. L., Diamond, M. C., Wu, S-Y., Slagle, R. W., & Saffran, E. Influences of environmental complexity and visual stimulation of occipital cortex in rat. *Brain Research,* 1969, vol. 14, 427-445.

Rosenzweig, M. R., Bennett, E. L., & Krech, D. Cerebral effects of environmental complexisty and training among adult rats. *Journal of Comparative and Physiological Psychology,* 1964, vol. 57, 438-439.

Rosenzweig, M. R., Krech, D., Bennett, E. L., & Diamond, M. C. Effects of environment complexity and training on brain chemistry and anatomy: a replication and extension. *Journal of Comparative and Physiological Psychology,* 1962, vol. 55, 429-437.

Rosenzweig, M. R., & Leiman, A. L. Brain function. *Annual Review of Psychology,* 1968, vol. 19, 55-98.

Rozin, P. Thiamine specific hunger. In C. Code (Ed.), *Handbook of physiology,* Sec. 6, Vol. 1, *Alimentary canal.* Washington, D.C.: American Physiological Society, 1967.

Rozin, P. Central or peripheral mediation of learning with long CS-US intervals in the feeding system. *Journal of Comparative and Physiological Psychology,* 1969, vol. 67, 421-429.

Rzóska, J. Bait shyness, a study in rat behavior. *British Journal Animal Behaviour,* 1953, vol. 1, 128-135.

Stanley, W. C., Cornwell, A. C., Poggiani, & Trattner, A. Conditioning in the neonatal pup. *Journal of Comparative and Physiological Psychology,* 1963, vol. 56, 211-214.

Teitelbaum, P., Cheng, M.-F., & Rozin, P. Development of feeding parallels its recovery after hypothalamic damage. *Journal of Comparative Physiological Psychology,* 1969, vol. 67, 430-441.

Thoman, E., & Arnold, W. J. Effects of incubator rearing with social deprivation on maternal behavior in rats. *Journal of Comparative and Physiological Psychology,* 1968, vol. 65, 441-446. (a).

Thoman, E. B., & Arnold, W. J. Incubator rearing of infant rats without the mother: Effects on adult emotionality and learning. *Developmental Psychobiology,* 1968, vol. 1, 219-222. (b).

Thoman, E. B., Wetzel, A., & Levine, S. Learning in the neonatal rat. *Animal Behaviour,* 1968, vol. 16, 54-57.

Wade, G. N., & Zucker, I. Development of hormonal control over food intake and body weight in female rats. *Journal of Comparative and Physiological Psychology,* 1970, vol. 70, 213-220.

Wiesner, B. P., & Sheard, N. M. *Maternal behaviour in the rat.* London: Oliver and Boyd, 1933.

## DISCUSSION

The discussion following Dr. Rosenzweig's presentation centered around two principal themes—some of the confusions and difficulties encountered in brain research, and some of the areas needing additional study.

In response to a question about the persistence of behavioral differences in experimental animals, Dr. Rosenzweig said that a number of investigators have shown that behavioral differences persist in rats which have been exposed to enriched environments and then returned to normal colony conditions. However, a great deal of additional investigation is needed. We are at the stage of knowing that there are cerebral effects but not knowing which variables produce them or what are the mechanisms by which they are produced. Even an hour of some types of manipulation produces cerebral effects, but other types of treatment do not produce changes.

It is not yet possible to say whether the behavioral effects so often reported with rodents are real differences of intellectual power or are simply differences in early adaptation to a testing situation. It has not yet been clearly shown that prolonged exposure to differential experience leads to differences in problem-solving ability that will persist over a long series of tests.

There are many areas of brain research that need further study, Dr. Rosenzweig continued. For example, little has been done to relate prior characteristics of rats to their behavior under certain test conditions. This might be a rewarding project. For example, one could characterize animals of a heterogeneous group as being high, medium, and low in their nonspecific excitatory level in relation to exploratory tendencies. These animals could be tested to see whether those that are more inquisitive and exploratory show larger brain weight effects of enriched environmental experience than do those that are low in exploratory behavior.

In brain studies, it is important to distinguish between experimental and learning effects, a participant said. We cannot always be sure that simply because we have affected an organism experimentally the effect has been mediated by learning circumstances. It is technologically possible to present the same stimuli in either a conditioning or a nonconditioning framework. In the group of animals exposed to stimuli in a conditioning framework, there might be facilitation which would be lacking in the group exposed to the same stimuli in a nonconditioning framework. Interpretation of data in brain studies is complicated further by the fact that differential experiences may, on the one hand, affect the brain structure of the young organism, or may, on the other hand, affect brain function. It is conceivable that experiential effects can be produced in the young organism that will facilitate or diminish his learning ability later on, even though the experience was not mediated by a learning process at the time. For example, animals nutritionally deprived early in their development later show poorer learning ability than animals who were not subject to nutritional deprivation early in life. This effect is not mediated by a learning process.

This is true in relation to nutrition, Dr. Rosenzweig agreed, but alterations in later learning behavior may also be affected by such emotional and motivational factors as alterations in the exploratory tendencies of the animals. The way animals transfer experience in various environments into problem solving with different formal bases is also under study. Exploratory tendencies and preferences for familiar or novel objects are being measured.

One problem is that the brain changes produced by formal learning in enriched environments tend to be quite small and difficult to identify and analyze. Often they are just tantalizing indications, and it is difficult to say positively that learning alone has brought about the changes.

The third leg of this particular stool is growth, another participant said. Perhaps we need to clarify what the interactions are among experience, learning, and growth.

The relationships is not always what might be assumed, Dr. Rosenzweig said. For example, sometimes the effect of enriched experience is to increase the ratio of cortex to the rest of the brain, which goes contrary to the usual pattern of development in the rat. Fairly large effects can be obtained even at ages where the growth of the rat's brain has reached a plateau. Thus it is evident that enriched experience does not simply accelerate or maximize growth.

In relation to growth and development, not all investigators agree on how early rats can be weaned without deleterious effects, another participant said. Czechoslovakian research indicates that rats weaned 14 to 18 days after birth show behavioral defects. Some researchers feel that weaning cannot be done before the age of 30 days without developmental defects, but an unpublished study by Levine, in which several behavioral measures were used, indicates no differences between rats weaned at 21 days and those weaned at 30 days.

For investigators interested in studying food mechanisms in young animals, the rabbit is a convenient research animal. In the natural setting, the mother rabbit builds a nest for her young in a hole in the ground. She feeds the young animals only once a day, at night, for a period of about 5 minutes. Then she goes about 1000 feet away and remains at a distance from them throughout the rest of the 24 hours. Thus any socialization of the young animals must be a function of peer group interactions rather than interaction with the mother, since she is almost never there. This pattern can be simulated in the laboratory by keeping the mother animal away from the young for all but 10 minutes out of the 24 hours. This bypasses one of the problems often encountered in nutritional studies which require young animals to be handled by investigators. If, after handling, the young animal is returned to the cage with his mother, she often picks him up and plays with him. Working with the rabbit eliminates this problem of technique, since there is no need to return the young animal to its mother after handling. This simple procedure has been reported in *Science* by Denenberg.

The interaction between heredity and environment is another area that is difficult to pin down, another participant said. No one doubts that a relationship exists, but it is very difficult to compute a specific number of genes necessary to produce certain traits. The single exception is of cases where the capacity for synthesizing an enzyme is missing, as in phenylketonuria. The notion of a one-to-one relationship between genes and traits is still in the realm of theory and vague concepts, even though we are on sound methodological ground in saying that there is, at all points, interaction between genes and environment.

The distinction between restructuring and regulation must also be kept in mind, the participant continued. In the embryonic state, animals can do something with their brains which in embryology is known as regulation. If a portion of the tissue is cut away, the animal will reorganize itself to establish exactly what was there before, even though there is less tissue and there is no new structure. But this is definitely regulation, not restructuring. Thus far it has not been possible to "hook up" the brain differently to force animals to create a brain structure or organization that was not there to begin with.

There are conflicting opinions about this, Dr. Rosenzweig replied. Some psychologists insist that the brain is quite plastic and its connections can be varied. However, Sperry's work indicates that the sensory and motor connections are very well formed by the time of birth, and that only minor changes in structure may occur with learning, perhaps in intermediate connections. The question of whether measurable gross anatomical changes take place in the brain with use remains unanswered.

Changes in brain structure are often looked at from too narrow a base, another participant said. It is fortunate that Dr. Rosenzweig has taken into consideration the increase in the number of glial cells in animals in an enriched environment. The tendency is often to think only in terms of neurons. Other structures such as histologic structure and vascular tissues also need to be considered. For example, it is conceivable that one primary effect of experimental factors might be mediated through changes in the vascular supply and the number and intimacy of glial cells. These may affect neuronal changes.

The whole area of glial work is growing, Dr. Rosenzweig responded. An article a few years ago contained 400 references selected from a much larger number that were available. "Glia" originally meant "glue" and implied that the glia pasted the brain together, but these cells are now known to serve many special functions. For example, they mediate between capillaries and neurons, and they provide the myelin sheathing. Different types of glia show different orders of change. Replication of studies in this area is needed. Some reports have been based on single experiments with very small numbers of animals, and in other studies there is great variability from experiment to experiment.

A participant inquired whether there have been studies of the brain structure of newborn rats whose mothers have had normal pregnancies, have been socially deprived during pregnancy, or have been socially stimulated during pregnancy. Is there evidence that stimulation of the mother during pregnancy affects the fetus?

A number of studies report that activity of the fetus is affected by the state of the mother, Dr. Rosenzweig replied. It may be just as Hippocrates and Harvey supposed, that the fetus gets stimulation and experience *in utero* and that this may be important in the organization of regulatory structures. This would be an interesting line of investigation to pursue more fully. Probably it would be preferable to use an animal better developed at birth than the rat, one with a longer intrauterine life span. The guinea pig, for example, has a longer gestational period and is born in a much more mature state than the rat. Thus the influences on the mother could operate over a longer period of time. The guinea pig brain would also be easier to work with than the very soft brain of a newborn rat. Some research has been done on the effects of the nutrition of the mother at different stages of pregnancy on the amounts of DNA and RNA in the brains of newborn rats.

There are many evidences of biochemical influences during intrauterine life, another participant said. Perhaps it would be productive to do histochemical studies of the brain of the newborn animal after the mother had been treated in different ways.

One reason for coming to conferences like this is the hope that one might interest someone else in taking up where his own work leaves off, Dr. Rosenzweig said. In brain research, far more needs to be done than can be accomplished by one laboratory. There is need for studies to follow up clues offered by current work, there is need for replication of studies, and there is need to do similar studies with different species of animals and, in some cases, with humans. The relationships between brain development and behavioral development are difficult to establish in an explicit, critical sense, but they are worthwhile areas for research.

# EFFECTS OF HANDLING IN INFANCY UPON ADULT BEHAVIOR AND ADRENOCORTICAL ACTIVITY: SUGGESTIONS FOR A NEUROENDOCRINE MECHANISM[1]

*Victor H. Denenberg and M. X. Zarrow*
University of Connecticut

Sufficient research has been generated over the past 15 years to clearly establish that various procedures of stimulating animals in infancy will result in fundamental changes in the adult behavioral and physiological capabilities of those animals. Initially, as with all scientific work, much of the research was concerned with documenting the presence of a reproducible phenomenon, in exploring various parameters of infantile stimulation, and in trying to determine which behavioral and physiological systems in adulthood were affected. We have now moved beyond this essentially descriptive stage and are presently concerned with investigating possible mechanisms of action at the physiological level. The purpose of this paper is to suggest that a neuroendocrine mechanism is centrally involved in mediating the effects of stimulation in infancy. Before getting to that section, however, it will be well to review briefly the behavioral work concerning infantile stimulation. This review will be restricted to work on rodents, primarily rats. In addition, the procedure called "handling" will be the only technique discussed here for stimulating animals in infancy. For a more general review of

[1] The research described in this paper was supported, in part, by Program Project Grant HD-02068 and HD-04639 from the National Institute of Child Health and Human Development, NIH.

the effects of infantile stimulation and early experience covering a variety of species see Denenberg (1969a).

## EFFECTS OF HANDLING IN INFANCY UPON LATER BEHAVIOR

When one looks back on the brief history of experimental research on early experiences, one finds that the first sets of studies were done using a postweaning animal in which the "richness" of the animal's environment was manipulated to see how this affected later perceptual and problem-solving behavior. This work was stimulated by Hebb's exciting book, *The Organization of Behavior,* published in 1949. It was not until several years later that experimenters turned to the newborn, or infant, animal for experimental investigations.

The first report concerning the effects of handling infant animals was probably the brief abstract by Hunt and Otis (1955) of a paper which they presented at the annual meeting of the American Psychological Association. Unfortunately, these researchers waited until 1963 before they published a formal report of their research (Hunt & Otis, 1963). The first full-fledged experimental research report on the effects of handling was a paper by Levine, Chevalier, and Korchin in 1956. In that study the authors were primarily interested in the effects of shock upon the infant animal's later ability to learn an avoidance response. However, being good experimenters, they were aware that their shocked group also had the experience of being removed from their home cages, being placed into a strange apparatus, etc., in addition to receiving the electric shock. Thus, they included a second experimental group which was manipulated in the same manner as their shocked group except that shock was omitted. This "handled" group was found to be as effective at avoidance learning as the animals in the shocked group, and both groups were better than a nondisturbed control.

In his next study, therefore, Levine (1956) omitted the shocked group and studied the effects of handling as an experimental variable. He had one group which was handled for the first 20 days of life, and a second experimental group which was handled at 50-70 days of age. The control group, of course, was not disturbed between birth and weaning, nor after weaning until avoidance testing began at 71 days. Levine found that the animals handled in infancy were significantly better in avoidance learning than the late-handled group or the control group, while the latter two groups did not differ from each other.

This experiment, in conjunction with the prior one by Levine *et al.* and the brief abstract by Hunt and Otis, established that handling, a seemingly mild form of stimulation in infancy, would have long-lasting effects upon the rat. In addition Levine's (1956) study also showed that handling in later life did not have the same consequences as handling in infancy.

*The Handling Procedure*

A word now about the technique which we use in handling our animals. First of all, to a naive observer the procedure we use appears to be very innocuous, and some people find it difficult to believe that such a simple procedure can have such long-lasting and profound effects. Fortunately, these effects have been well documented so that it is no longer necessary to convince people of the realities of one's research findings. In our laboratories pregnant rats are housed in stainless steel cages with a tray floor covered with shavings. A food hopper and a water bottle are attached to the outside of the front door of the cage so that the animals can be fed and watered without having to open the cage door, thus minimizing disturbance to the animal and her litter.

The cages are checked every morning, and when a newborn litter is discovered the tray containing the young is removed from the cage, leaving the mother in the cage. The young are sexed at this time and the litter is reduced to eight pups of prescribed sex ratio. If the litter has been randomly assigned to the nonhandled control condition, the tray is immediately replaced in the maternity cage with the eight young, and neither the mother nor young is disturbed until they are weaned at 21 days. As mentioned above, food and water are supplied through external sources so the cage door does not have to be opened. Also, the shavings are never changed during the nursing period.

If the litter has been assigned to be handled, then each of the eight pups is placed separately into a tin can containing shavings. The pups are left alone in the can for 3 minutes. They are then placed back onto the tray, and the tray is returned to the maternity cage. The above may continue daily throughout the first 20 days of life (we routinely wean our animals at 21 days) or the animals may be handled for a lesser number of days. At no time during the manipulation of the young pups are they fondled or caressed. Thus, when we speak of "handling," we refer to the procedure described above, and not to any manual manipulation of the young animal. In a sense, the word "handling" is not a good one to use because it is not an accurate description of the procedures involved. However, no one has been able to think of a better term, so we use this one.

As is to be expected, not everyone uses the same procedure in the handling of their rats or mice. For example, in studies where electric shock is used, the "handled" group is often one which is placed upon the grids but does not receive shock, as was done in the Levine *et al.* study described above. This is clearly a very different procedure than placing animals into a tin can containing shavings because in the latter situation there is more conservation of body heat than there is when an animal is placed on grid bars and is exposed to surrounding air currents.

*Research Findings*

And so with the demonstration that handling was an effective way to manipulate the animal during its very earliest postnatal days, a number of researchers started looking at various behavioral endpoints which were modified by handling. A partial list of behaviors which were affected by handling stimulation includes: open-field activity and defecation, consummatory behavior, various timidity tests, avoidance learning, discrimination learning, exploratory behavior, aggression, and maternal behavior. Physiological endpoints which were shown to be affected include: body weight, sexual precocity, resistance to leukemia virus, ascorbic acid depletion, and ability to survive terminal stress. This material is reviewed in several places (e.g., Denenberg, 1967, 1969a; Levine, 1962b; Newton & Levine, 1968).

One of the important generalities that has emerged from this research was that animals which had been handled in infancy were less "emotionally reactive" than were animals from the nonhandled control group. Emotional reactivity is a construct, and different experimenters had used many different behavioral tests to tap part of this construct, including activity and defecation in the open field, consummatory behavior, ratings of emotional reactivity, performance in a learning situation where there is noxious reinforcement, and various measures of timidity. The rationale underlying all of these tests is that the emotional animal will freeze and defecate while the nonemotional animal will tend to explore his environment (Denenberg, 1967, 1969b). Denenberg (1964) attempted to integrate these various findings concerning infantile stimulation and emotionality in a theoretical paper. The central hypothesis of that paper was that emotional reactivity is reduced as a monotonic function of amount of stimulus input in infancy.

This hypothesis has been of heuristic value in suggesting several research ideas. One of these ideas comes from an assumption made in Denenberg's 1964 paper. This assumption is that the various tests which were purported to be behavioral measures of the construct of emotionality were indeed measuring that construct. In order to test that assumption Whimbey and Denenberg (1967a) generated 16 experimental groups by manipulating four variables during infancy and early postweaning life. The nature of the experimental variables is not relevant to this discussion except to note that two of them involved handling: we had two populations of mothers, one of which had been handled in their infancy while the other group had not. In addition, half of the litters from each of these two groups were themselves handled in infancy. These 16 groups of animals were given a large variety of behavioral tests when they were adults, including a number of purported measures of emotionality. These were open-field activity and defecation, a rating of emotional reactivity when the animal was removed from his home cage, consummatory behavior following stress, a measure of defecation in an avoidance conditioning apparatus which the animals had learned to fear, and activity in an apparatus containing noxious stimuli. These data were subjected to a factor analysis to test the hypothesis that these several measures of emotional reactivity would emerge in the same factor. They did. In addition, we were able to isolate a second factor, orthogonal to the emotional reactivity factor, which dealt with activity in various apparatuses and which we named "field exploration." In other words, emotionality and exploration are not bipolar opposites of one dimension but are, instead, two independent behavioral dimensions.

The detailed analysis of the activity and defecation scores in the open field taken over several days of testing revealed that the defecation measure loaded heavily on the emotionality factor while the activity score was factorially complex, loading both on emotionality and on field exploration. To make matters even more complicated, activity on the first day in the apparatus was *positively* correlated with emotional upset while the typical negative correlations were obtained on all subsequent days (Denenberg, 1969b; Whimbey and Denenberg, 1967b).

To summarize this section, the technique of handling rats between birth and weaning (and again it is important to stress the operational definition of handling) results in a wide range of changes in the animal's adult behavioral and physiological capabilities. One major generalization which may be made is that the emotional reactivity, or emotionality, of the rat is markedly and significantly reduced by this early handling experiences as measured in a variety of behavioral test apparatuses.

Having demonstrated behavioral differences in emotionality as a function of handling in infancy, the natural question that arises is whether there are any correlated changes at the physiological level. Changes in emotional reactivity would seem to implicate the adrenal gland, and, ultimately, the complete hypothalamic–pituitary–adrenal axis. This leads us to the next part of this article, which involves the hormone, corticosterone, secreted by the adrenal cortex.

## EFFECTS OF HANDLING IN INFANCY UPON THE ADRENAL AND THE PLASMA CORTICOSTERONE RESPONSE AND EMOTIONALITY

In the late 1950 s Levine became interested in the physiological reactivity of rats which had been stimulated in infancy and published several papers on this general topic. A review may be found in Levine (1962b). The first study measuring plasma corticosterone as a function of stimulation in infancy was done by Levine (1962a). The value of measuring corticosterone, as compared to other indices of adrenal activity (e.g., adrenal ascorbic acid depletion or adrenal weights), is that corticosterone is the major hormone secreted by the adrenal cortex of the rat and is specifically released as a function of ACTH from the pituitary gland. The release of corticosterone occurs whenever the animal is exposed to a novel or noxious stimulus situation, and the hormone is involved as a part of the homeostatic response to an emergency situation. Other indices of adrenal activity do not have this high degree of specificity, so there is ambiguity in interpreting the results obtained with those measures. It should be noted that the measurement of corticosterone tells us nothing about the activity of the adrenal medulla and the circulating catecholamines.

In Levine's study (1962a) there were handled and nonhandled animals. In adulthood some animals from both groups were removed from their cage, decapitated immediately, and blood collected. This was done to obtain a resting level of corticosterone prior to the introduction of any noxious stimulus. Other animals were subjected to a series of electric shocks and were decapitated at varying times after the cessation of the shock and their blood assayed for corticosterone. Levine found that the handled animals responded more rapidly with an increase in corticosterone than did the nonhandled controls, and that the handled group retained the higher level of corticosterone output throughout the 15-minute period of the experiment.

Several years later we obtained data which were, in general, confirmatory of Levine's findings (Haltmeyer, Denenberg, & Zarrow, 1967). We found that, for the first 5 minutes after the termination of electric shock stimulation, handled animals had a higher amount of corticosterone in their blood than did nonhandled controls, but we also found that the curves crossed over at 15 minutes with the controls having a somewhat higher amount than the handled animals.

The data of these two experiments appear to present a paradox. We had previously stated that handled animals were less emotional as measured by various behavioral tests than were nonhandled controls. A parallel finding at the physiological level would be that handled animals would give a lesser corticosterone response when placed into a strange situation, but we reported above that the handled animals gave a greater corticosterone response than controls to the stimulus of electric shock. However, there is an obvious variable which may be contaminating the results: The various behavioral tests used to measure emotional reactivity did not use a stimulus as stressful as electric shock. The typical procedure was to place animals into a strange or novel situation and observe their behavior in this setting. Also, to talk about corticosterone and emotionality measures in the same breath was premature since no one had obtained measures of plasma corticosterone and measures of emotional reactivity on the same group of animals. And so in 1967 Levine, Haltmeyer, Karas, and Denenberg carried out and reported a study which investigated several of these conditions.

Both handled and nonhandled animals were used, with 144 rats in each group. All animals were tested in the open field. At the end of the first day's testing 36 handled and 36 nonhandled animals were decapitated 0, 5, or 15 minutes after the end of open-field testing ($N = 12$ per group), and their blood was assayed for plasma corticosterone. The remaining 108 animals in each group were tested in the open field the next day, and again 36 animals from each group were decapitated at the termination of that day's testing. This same procedure continued for four successive days. The results of this experiment are shown in Figures 1 and 2.

On all 4 test days the handled group had a lower percentage of animals defecating than the nonhandled controls. With respect to the activity data, the nonhandled controls were more active on Day 1 (recall our discussion concerning the factor analysis findings in which Day 1 activity scores were found to correlate positively with the factor of emotional reactivity), while they were lower on the subsequent 3 days. Figure 2 presents the corticosterone data and shows that the handled animals were less reactive physiologically than nonhandled controls. The importance of this experiment is twofold: first, in demonstrating that under conditions of the mild stress of the open field the handled animals are physiologically less reactive than controls, and second, in demonstrating a correlation between the behavioral measures of emotionality and the physiological indicator of adrenocortical activity.

In all the studies mentioned above the animals had been handled for 20 days, which is the complete period of infancy. A meaningful question which arose was whether one could show a "dose-response" curve as a function of the amount of handling in infancy. You will recall that Denenberg (1964) had hypothesized that emotional reactivity is reduced as a monotonic function of amount of handling in infancy. The previously discussed study by Levine *et al.* (1967) had shown a significant relationship between adrenocortical activity and open-field performance so that the corticosterone response could be looked upon as a possible physiological measure of emotional reactivity. Denenberg and Haltmeyer (1967) investigated this question by handling rats for 0 days (i.e., nonhandled controls), 5 days, 10 days, or 20 days in infancy. The 5-day group contained animals which had been handled on Days 1-5, 6-10, 11-15, and 16-20; the 10-day groups contained animals handled on Days 1-10 and 11-20. These various groups were averaged, in this way balancing out the period during infancy when handling took place. At 21 days a number of these animals were placed into a novel environment where they remained for 10 minutes, at which time they were removed, decapitated, and their blood assayed for corticosterone. We found a linear dose-response relationship between amount of handling and corticosterone with the group that had received no handling experience giving the greatest corticosterone response while the group which had been handled for 20 days gave the least response.

One final study will round out this section. In all the experiments described above, the animals had been left in the novel or noxious environment until they were removed and killed. We had consistently found under these conditions that there was no evidence of habituation or adaptation as measured by cortico-

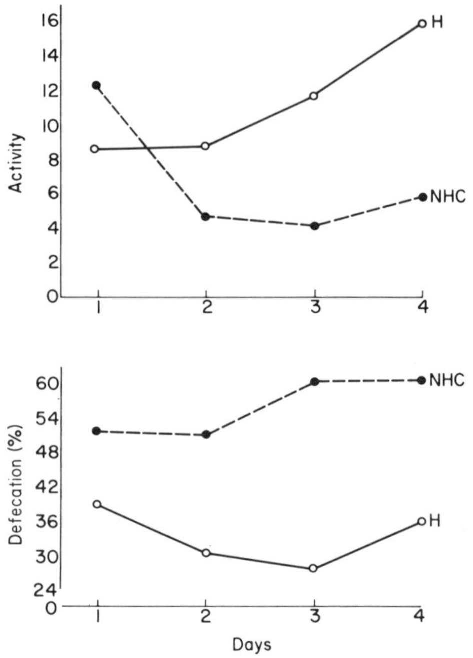


**Fig. 1.** Open-field activity and defecation for handled and nonhandled rats over 4 days in the open field.

sterone secretion. However, in studying a physiological process, such as the corticosterone response, it is important to find a technique which will allow one to measure the onset, magnitude, and duration of the response until it returns to its resting level again. We were able to develop such a technique by placing the animals into a novel environment for a brief period of time and then returning them to their home cage—a familiar environment. We left them in their home cage for varying periods of time and then removed them, decapitated them immediately, and assayed their blood for corticosterone. Figure 3 shows the results of such an experiment (Hess, Denenberg, Zarrow, & Pfeifer, 1969).

Figure 3 shows that the nonhandled controls give the greatest corticosterone response to the transient exposure to the novel stimulus situation. They maintained this response for a greater period of time than handled animals, and at the end of 1 hour they had almost returned to their baseline. The animals handled for 20 days showed the least increase in adrenocortical activity as a function of exposure to the open field, and they had fully recovered at the end of the hour. The two groups handled on the first 5 days of life only (for purposes of this discussion the two temperature conditions are not relevant) had scores intermediate between the nonhandled controls and those handled for 20

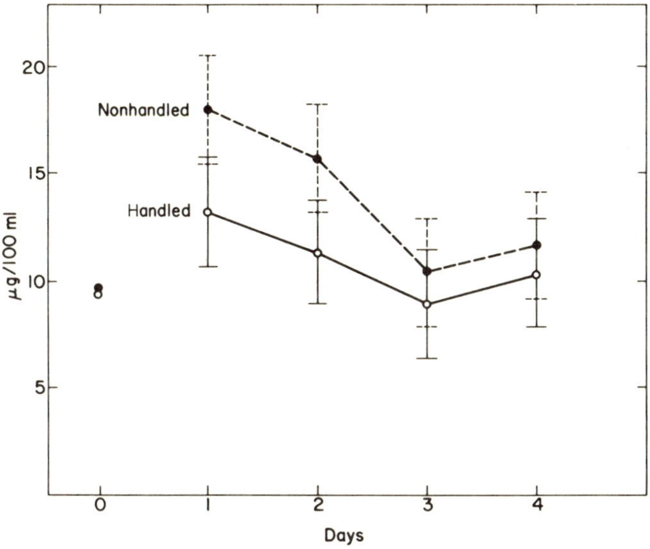


**Fig. 2.** Plasma corticoid levels for handled and nonhandled rats over days following 3 minutes in the open field. The points represent the mean value and bars represent the standard error of the mean. The two points at 0 days are the base levels of corticosteroids for handled and nonhandled rats. From Levine *et al.* (1967).

days. These data are also consistent with our previous findings of a dose-response relationship in that the nonhandled group had the greatest response, those handled for 5 days were next in magnitude of response, and those handled for 20 days had the least response.

In summary, we have been able to demonstrate that animals handled in infancy will give a significantly lesser adrenocortical response to mild novel stimuli than will nonhandled controls, while they will give a greater corticosterone response to a distinctly noxious stimulus like electric shock. These findings suggest that the handled animal is more adaptive in his physiological reactivity to stressful stimuli. Levine (1962b) has suggested essentially the same conclusion. We have also demonstrated that a relationship exists between behavioral measures of emotional reactivity and the response of the adrenal cortex.

The data from these several experiments are what led us to the working hypothesis that corticosterone might be involved in mediating the effects of infantile stimulation. That hypothesis was based on other recent data concerning the effects of the sex hormones acting prenatally or in early infancy upon brain organization. We used the neonatal sex hormone data as a model for our ideas about corticosterone. Therefore, it is necessary to move off at a tangent for a few moments to describe briefly the sex hormone data.

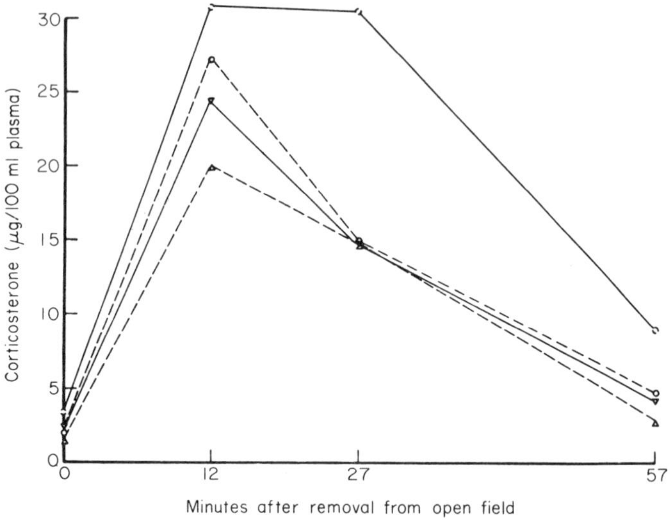


**Fig. 3.** Time course of the plasma corticosterone response in early adulthood as a function of handling and temperature conditions in infancy. ○—○, Control; ○--○, handled days 1-5, 35.5 °C; △—△, handled days 1-5, 22.2 °C; △--△, handled days 1-20, 22.2 °C. From Hess *et al.* (1969).

## NEONATAL SEX HORMONES AND BEHAVIOR: A MODEL FOR INFANTILE STIMULATION

The sex hormones have two different functions that are related to the age of the animal. Either prenatally or shortly after birth (depending upon whether or not the animal is precocial) the role of the sex hormones is that of *organizing* the central nervous system so that the animal will have either a male or a female brain. In adulthood the function of the sex hormones is that of *activating* an already organized nervous system so that appropriate sexual behavior takes place. Reviews of this material may be found in Young, Goy, and Phoenix (1964), Levine (1966), and Levine and Mullins (1966). This summary will merely cover the principles involved in order to establish the model.

Using the rat as the experimental animal, one can show that if the male is castrated within the first few days of life, and in adulthood is given estrogen and progesterone (i.e., the female hormones), such an animal will exhibit complete female sexual behavior. This will not occur if the male is castrated as an adult and given female hormones. Now consider the fairer sex. If a female is injected with testosterone in early infancy, and reinjected with the male hormone in adulthood, she will exhibit masculine behavior. However, the injection of testosterone into a normal adult female rat will not have this masculinizing effect. Consider also the presence of cyclic sexual events in the female and the lack of cyclicity in the male. Such cyclicity is still present in the female even if castrated within the first few days of life. The male will also exhibit cyclicity if castrated during the neonatal period. Adult cyclicity in both sexes will be prevented by the presence of testosterone (either via injection or by the testicles) in early infancy.

From these as well as other results the generalization has been drawn that the basic behavior pattern of the mammal is feminine and that a masculine pattern occurs if the hormone, testosterone, is present in the newborn. A number of studies, including those cited above, have shown that it is necessary for these hormones to be present in the rat within the first 5 or so days of life to exert their influences. Evidence that this effect is at the level of the central nervous system has been shown by experiments in which male and female pituitaries have been transplanted without having a differential effect, and by other experiments in which sex hormones have been placed directly into the brain where they caused appropriate sexual events to occur.

In summary, the rat's adult sexual behavior is determined within the first 5 days of life by the presence or absence of a hormone which acts upon the central nervous system, presumably the hypothalamus, and establishes a neural pattern there which determines whether the animal will act in a masculine or feminine fashion throughout its adult life.

## STUDIES INVOLVING CORTICOSTERONE AS PART OF THE MECHANISM OF INFANTILE STIMULATION

One of the very important principles coming out of the work on neonatal sex hormones is that hormones either prenatally or in early infancy can act upon and modify brain organization. Though this has just been established for the sex hormones, it seems reasonable to conclude that other hormones may also have this function. When one now looks at the data on infantile stimulation and compares them with the neonatal sex hormone data, certain interesting parallels are seen. First of all, the effects of infantile stimulation are limited to the period of infancy, and the sex hormones also have their impact during very early infancy. [Remember that Levine (1956) showed that handling between 50-70 days of age had no effect upon behavior as compared to handling between 1-20 days of age.] Also, most of the effects of handling during the first 20 days of life can be obtained by handling animals for the first 5 days (see the two 5-day groups in Figure 3 and compare them with the 20-day group). Second, handling has long-term essentially permanent effects on the behavior of an animal, and the same is true with the neonatal sex hormones. (This statement does not imply that it is impossible to reverse these effects. It merely states that given a constant set of conditions under which animals are maintained, one is able to demonstrate these effects throughout the lifetime of the animal.) Finally, corticosterone is a steroid and, as such, belongs to the same class of hormones as testosterone, estrogen, and progesterone.

Because of these various considerations, we adopted the working hypothesis that stimulation in infancy brought about the release of corticosterone into the bloodstream, and that the corticosterone went to the brain, presumably to the hypothalamus, where it acted to modify brain organization, thus resulting in an animal which was less emotionally reactive. We set out to test this hypothesis in a series of experiments which we will now briefly describe.

If our hypothesis is true, the first thing which must be established is that corticosterone can be found in the blood of a neonatal animal. Interestingly enough this had not been demonstrated previously and, indeed, there was some evidence that corticosterone was not present. We will get to that in a moment.

In considering how to attack this problem we decided that the safest thing to do was to use very powerful stressors to try to elicit the corticosterone response. Therefore we selected electric shock and heat. We used 0.8 mA of shock delivered to the young animal on a brass grid floor. For heat stress the animals were suspended in a nylon mesh basket in a thermal-regulated chamber maintained at 63°C. The animals were exposed to these stressors for 3 minutes. They were then removed, placed in a can containing shavings for an additional 15 minutes, after which they were decapitated, and both plasma and adrenal corticosterone were determined. Since we had no knowledge as to the age at which the animals would be sensitive to these stressors, if at all, we decided to do a developmental study. Different groups of animals were stressed and assayed at 1, 2, 3, 4, 5, 7, 9, 11, and 21 days of age (Haltmeyer, Denenberg, Thatcher, & Zarrow, 1966). Figure 4 shows the data for the amount of corticosterone in the plasma, while Figure 5 presents the findings for the amount of corticosterone obtained in the adrenal cortex itself.

The plasma corticosterone data in Figure 4 clearly demonstrate that the newborn rat is capable of secreting significant amounts of corticosterone into its bloodstream as a function of exposure to electric shock and heat stress, with heat having a greater effect than shock during the first 5 days of life. Another interesting thing to note about the curves in Figure 4 is that there is a considerable amount of corticosterone present at birth, that the levels drop to a minimum at approximately 7 days of age, and then rise again at 21 days for the experimental animals to values approximating those of the normal adult. This is an important point to which we will return later in the paper. Figure 5, which shows the amount of corticosterone which can be measured in the adrenal gland, presents a slightly different picture. Heat has a measurable effect starting with the first day of life, but electric shock had no significant effect upon increasing corticosterone output until the animals were 7 days old.

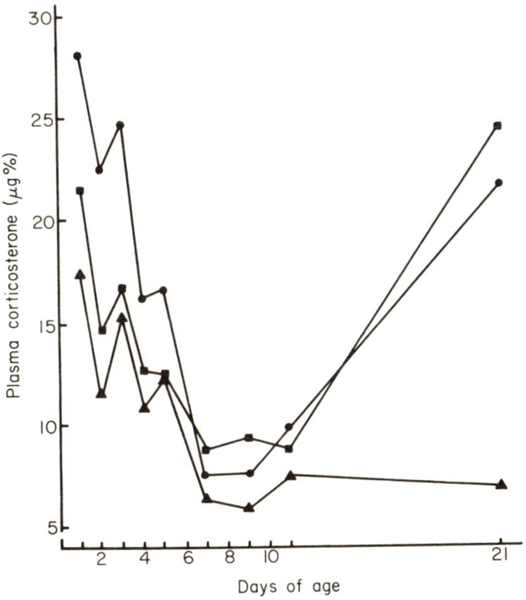


**Fig. 4.** Plasma corticosterone concentrations in the immature rat following exposure to heat stress, electric shock, or no stimulation. ■, Shock; ●, heat; ▲, control. From Haltmeyer *et al.* (1966).

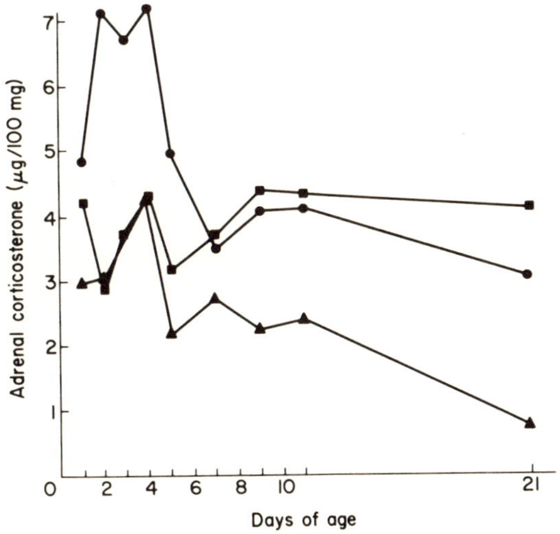


**Fig. 5.** Adrenal corticosterone concentrations in the immature rat following exposure to heat stress, electric shock, or no stimulation. ■, Shock; ●, heat; ▲, control. From Haltmeyer *et al.* (1966).

These results were consistent with our working hypothesis, which gratified us. However, the results were contradictory to other findings in the literature, and this perplexed us. Several researchers had reported over the years that the neonatal rat did not respond to stressors with any evidence of an adrenal reaction. In fact, Schapiro and his associates (Schapiro, 1962; Schapiro, Geller, & Eiduson, 1962) had formalized this observation by coining a term for it. They stated that the rat had a "stress-nonresponsive period" during the first 6 to 8 days of life during which time stressors would not produce any adrenal activity.

We were now faced with the question as to why we had found evidence of significant adrenal activity while Schapiro and his associates had not. To investigate this discrepancy in findings we carried out a second experiment using rats which received heat or electric shock stress at 2 days when, according to Schapiro, they are nonresponsive and at 9 days when they are known to be responsive (Zarrow, Haltmeyer, Denenberg, & Thatcher, 1966). At both 2 and 9 days the animals were subjected to the same electric shock or heat stress for a period of 3 minutes as had been described previously. They were then placed into shaving cans and were killed at varying times thereafter. In order to get resting levels, some animals were killed as soon as they were removed from the home cage. Figures 6 and 7 show the data for plasma corticosterone and adrenal corticosterone, respectively, in this experiment.

As compared to resting level values (designated by the letters RL in Figures 6 and 7) the plasma corticosterone response of those animals which received either heat or electric shock stimulation at 2 and 9 days of age was significantly increased, which is consistent with our prior findings. Also in Figure 6, it will be noted that there is a greater response at 2 days of age than at 9 days of age, again consistent with our previous study. The discrepancy between our results and those of Schapiro and his associates was explained by the adrenal corticosterone data. In the left-hand panel of Figure 7 we see that electric shock stimulation has no measurable effect upon the amount of corticosterone found in the adrenal gland. Thus both we and Schapiro failed to find any evidence of an adrenal response when 2-day old rats were stressed with electric shock, killed an hour afterwards, and assayed for adrenal corticosterone. If the first or third of these parameters is changed, significant effects are obtained. Thus, the stress-non-responsive hypothesis was based upon an artifact generated by selecting an unfortunate combination of parameters to investigate this phenomenon. In Figure 7 we again found evidence confirming our previous study that heat has a very marked effect both at 2 days and 9 days of age, and that electric shock also has a significant effect at 9 days.

Our first two experiments have established that the newborn rat could respond to stressors such as heat and shock with a significant rise in adrenal corticosterone. Our next concern was whether other forms of stressing agents could also elicit a significant increase in adrenal activity in the 2-day old rat

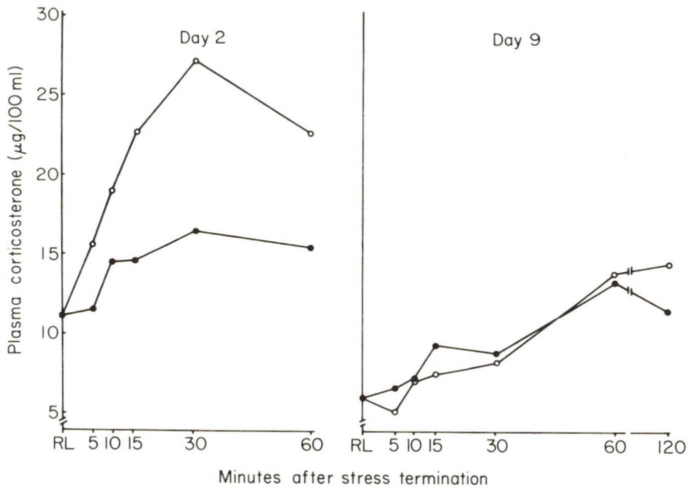


**Fig. 6.** Time course of plasma corticosterone following exposure to heat (○–○) and electric shock (●–●) in the 2-day and 9-day old rat. From Zarrow *et al.* (1966).

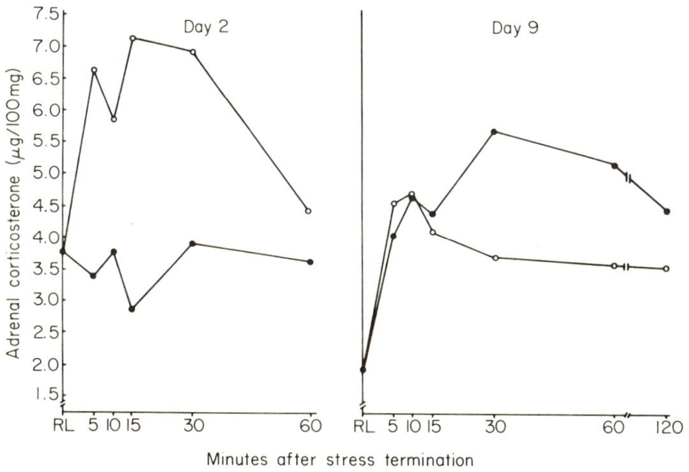


**Fig. 7.** Time course of adrenal corticosterone following exposure to heat (○–○) and electric shock (●–●) in the 2-day and 9-day old rat. From Zarrow *et al.* (1966).

(Zarrow, Denenberg, Haltmeyer, & Brumaghim, 1967). When ACTH and histamine were used as stressors, significant rises in corticosterone were obtained. When the animals were exposed to cold stress by being placed on crushed ice, only a marginal effect was obtained.

There was one stressor in this experiment which did not have any measurable effect, namely ligation: surgical silk was tied tightly just above the knee of the right hind leg and the ligature was tight enough to cause the animals to squeal. Yet in spite of this squealing, which we take to be evidence of a painful stimulus, we found no evidence of an increase in either plasma or adrenal corticosterone as compared to controls. We consider this finding of the ligature experiment to have rather broad implications. Fortier (1951, 1966) has suggested two classes of stressors, one "neural" and the other "systemic" according to whether they activate ACTH release solely by the mediation of the nervous system or whether they act on the pituitary through the systemic circulation. The outstanding feature of the stimulus involved in the ligature experiment appears to be its neural nature (the ligature was not tied tightly enough to cut off blood circulation in the limb). These results would tend to indicate that the pituitary–adrenal axis is operative in the 2-day old rat but that the neural component in the system is not active as yet. This finding would also suggest why we found a lesser effect with electric shock in our first experiment of this series (see Figures 4 and 5) than with heat since electric shock has a large neural component involved while heat would have its effects mediated primarily through systemic activity.

These findings had started us off on the right track by showing that it was possible to get the adrenal gland to release corticosterone into the blood stream when strong stressing agents such as heat, shock, histamine, or ACTH were used. But the technique we had used to stimulate infant rats in our early experience work was the handling procedure which certainly appears to be much more innocuous than the various stressors enumerated above. The experimental question to which we next addressed ourselves was whether the standard handling procedure was a sufficient stimulus to elicit a measurable corticosterone response in the neonate (Denenberg, Brumaghim, Haltmeyer, & Zarrow, 1967).

On the first day of life litters were cut back to eight animals, but they were not handled at this age because we wanted to keep this experiment comparable to our previous studies. On Day 2 the pups were removed from their maternity cage, two were killed immediately from each litter to obtain resting levels, and the other six were placed into handling cans where they were left for 3 minutes. At the end of the 3 minutes they were placed back with their mother in the maternity cage. This procedure, it will be noted, is identical to our standard handling techniques except that six rather than eight animals were returned to the mother. Thirty minutes after being returned, two pups were removed from the litter and were killed. We found that those pups which had received handling had a significantly greater amount of plasma corticosterone than their resting level littermate controls.

We also found two other techniques which would elicit a significant adrenocortical response. The first was to leave the pups along in handling cans for 33 minutes, and the second was to place the pups into an empty maternity cage for a half hour after the usual 3 minutes in handling cans. These two procedures demonstrated that it was not necessary to return the young to the mother in order to obtain a significant response. The importance of this was in eliminating the mother as necessarily a causal variable. The mother, upon the return of the pups, will pick the young up, run around with them, prepare a new nest for them, and engage in much "handling" activity with them. Thus, it is possible that it is the mother's behavior toward the pups after they are returned to the nest box which is the major stimulus that affects the pups' subsequent behavior. (We are in agreement that the mother's behavior toward the pups does have a significant effect, but it is also important to understand clearly that the handling procedure, independent of the mother's behavior, can also elicit a significant corticosterone response.)

Another part of this study investigated the question as to whether the temperature loss brought about as a function of being removed from the warm nest and being placed into a handling can alone was a significant factor in eliciting the corticosterone response. If temperature loss was the critical variable, then the corticosterone response should be prevented if rats were placed into handling cans maintained at a temperature equivalent to that present when the rats were packed together in the maternity cage. The average temperature in a packed nest was 35.5°C. When 2-day old pups were placed into handling cans maintained at this temperature, their corticosterone levels did not differ from that of resting level controls, while other experimental animals placed in handling cans at ambient temperature did show the usual significant rise in plasma corticosterone.

But one should not suppose from this that the change in temperature is the only variable affecting the corticosterone response. In Figure 3 there was one group of animals which had been handled for 5 days in infancy at body temperature, and this group also had a significantly lesser corticosterone response to the novel stimulus of an open field than did nonhandled controls. Thus one is able to handle animals without reducing their body temperature and can still obtain significant effects. We suggest that these significant effects may be mediated through the mother's behavior toward the pups after they are returned to the nest box.

Let's pause for a moment and recall our working hypothesis which set us off on this series of studies. It was that stimulation in infancy would bring about a release of corticosterone from the adrenal cortex into the blood stream and that this hormone would act upon the central nervous system, presumably the hypothalamus, in order to change the neural organization of the brain. We have gotten corticosterone into the blood stream by our handling procedure as well as by other more severe stressing agents. The next question was: where does the corticosterone go from there? To answer that question we took radioactive-labeled corticosterone and injected it into 2-day old rats. Different groups were killed 10, 30, 60, or 120 minutes after the injection, and we removed muscle tissue, and liver, the hypothalamus, and the rest of the brain, and we counted the amount of corticosterone present using a liquid scintillation system. The radioactive corticosterone was taken up in significant amounts both by the hypothalamus and by the rest of the brain throughout the 2-hour interval of the study. More important, however, was that the hypothalamus showed a greater uptake of radioactivity than the rest of the brain both at 10 minutes and 30 minutes after injection (Zarrow, Philpott, Denenberg, & O'Connor, 1968). At the same time McEwen, Weiss, and Schwartz (1968) showed a selective retention of corticosterone by the hippocampus and septum.

Now let us change focus for a moment. The technique of using radioactive-labeled corticosterone had been useful in demonstrating that this hormone is taken up by the brain, with a greater amount in the hypothalamus. Recall that in all the experiments described above our various manipulations had been mediated by the experimenter. However, in nature no experimenter is present, and so if we believe that corticosterone is having an effect upon the newborn animal in its natural setting and if we believe that handling has some counterpart in a biological context, we have to find some natural biological mediator for these events. It seems reasonable to assume that the mother would be the mediator during both the prenatal stage and the postnatal stage of development.

Consider, first, prenatal development. A number of studies have demonstrated that a pregnant rat which is emotionally upset will have offspring which are also emotionally upset (Ader and Belfer, 1962; Thompson, 1957). On the other hand, Ader and Conklin (1963) demonstrated that when pregnant rats are handled during their gestation, the offspring are less emotional. If corticosterone from the pregnant mother is able to pass the placental barrier, it is possible that this hormone might act upon the central nervous system of the developing fetus to affect its neural organization and, thus, its emotional behavior. To test this hypothesis, we injected radioactive corticosterone into pregnant rats the day before they were due to deliver their young (Zarrow, Philpott, & Denenberg, 1970a). Thirty minutes later the mothers were decapitated, the fetuses were removed, and tissues from muscle, liver, hypothalamus, and the rest of the brain were evaluated to determine that amount of radioactivity present in the fetuses. We found that the hypothalamus and the rest of the brain had taken up a significant amount of radioactive corticosterone and again, as in our prior experiment, the hypothalamus had a significantly greater uptake than did the rest of the brain. Thus maternal corticosterone is able to cross the placental barrier and can localize in the brain of the fetus, thereby indicating that this is a possible mechanism for the effects of prenatal stimulation.

The next question we asked was whether corticosterone could get through the mammary glands of the mother while the young were nursing (Zarrow, Philpott, & Denenberg, 1970a). On the second day after birth the mothers were separated from their pups for 4 hours and were then injected with radioactive corticosterone. The pups were then returned and allowed to nurse for 55 minutes. We were able to find evidence of a measurable amount of radioactive corticosterone in the blood of the neonatal animals and in their stomach contents, but in no other tissues. Furthermore, only a very small percentage of the injected hormone was found in the offspring, which agrees with findings of others that labeled steroids are transferred through the milk in only extremely small amounts. Thus, it seems unlikely that corticosterone in the mother's milk has much effect upon the infant animal's behavior. It is much more likely that the mother affects her offspring by means of her behavioral interactions with them rather than through her milk supply. Evidence supporting this position has been recently reviewed by Denenberg (1970) with respect to the effects of maternal factors upon the aggressive behavior, open-field activity, and adrenocorticosterone response of mice.

This brings us up to date with respect to our research on the hypothesis that corticosterone is involved in mediating the effects of infantile stimulation. So far each experiment that we have done has been completely consistent with our hypothesis, though we recognize that this does not establish its validity. To summarize briefly, we have shown that corticosterone is released in the neonatal rat as a function of intense stressors such as shock, heat, histamine, and ACTH. We have also shown that our standard handling procedure will bring about a significant release of corticosterone in the 2-day old animal. We have been able to obtain a corticosterone response in the newborn animal without returning it to its mother, thereby demonstrating that we can get an effect mediated entirely by the experimenter. This does not mean that the mother has no effect. Instead, it suggests that our handling procedure may do some of the same things that the mother does toward her young between birth and weaning, as if we are, in a sense, simulating certain characteristics of the mother's behavior. McEwen *et al.* (1968) have shown a retention of corticosterone by the hippocampus and septum, and we have been able to demonstrate that the corticosterone, after it gets into the blood stream, goes to the brain and more of it is taken up in the hypothalamus than in the rest of the brain. Our results indicate that this is true for the newborn animal and is also true for the fetus, thus suggesting a mechanism whereby emotional upset on the part of the mother during pregnancy may have an effect upon her unborn offspring's subsequent emotional behavior. When we look at the nursing female, we find evidence that only a very small amount of corticosterone in the milk supply has any major effect upon the animal's behavior. Instead, we think it more reasonable that the mother's behavior toward her young results in stimulation that brings about a release of endogeneous corticosterone.

Now that we have established that corticosterone does go to the brain, one of the next steps in our research program is to find whether it has any effect on modifying the neural organization of the hypothalamus or the rest of the brain. That problem should keep us occupied for the next several years, easily.

## STUDIES ON THE PITUITARY–ADRENAL SYSTEM OF THE NEONATAL RAT

Our working hypothesis is that the release of corticosterone as a function of stimulation in infancy is what initiates the chain of events resulting in subsequent behavioral and physiological changes of the animal. A thorough test of this hypothesis demands an understanding of the physiological activity of the neonatal adrenal cortex and its relationship to the pituitary gland. Thus, we have been carrying out a series of studies concerned with determining what is happening to the adrenal cortex of the rat during its first 21 days of life, and how this is related to pituitary activity. Data on this topic were presented in Figures 4 and 5 which dealt with the developmental curves of corticosterone release for control animals and those stimulated by electric shock or heat. One interesting characteristic of those curves is that there is a high level of corticosterone output within the first few days after birth, falling to a low level at around 7-9 days, and then returning to a high level for our experimental animals at 21 days. A similar pattern has been found for many other aspects of adrenal physiology. One such characteristic is shown in Figure 8, which plots the relative weight of the adrenal gland as a function of age between 2 days and 30 days. The relative weight drops to a minimum at 7 days of age and slowly recovers up to 15 days of age after which it becomes asymptotic. The relative weight is, of course, the ratio between the total body weight of the animal and the total weight of the adrenal gland. What this curve depicts is that the overall body weight of the animal grows quite rapidly throughout the 30-day period while there is very little growth with respect to the adrenal gland between 2 and 7 days of age and then it begins to grow at an increasing rate (Zarrow, Philpott, & Denenberg, 1970b). Thus, the atrophy seen in Figure 8 is more apparent than real.

But we have found evidence that this "biphasic" curve is a real phenomenon. In the first of two experiments, Zarrow, Philpott and Denenberg (1968) removed the right adrenal gland of rats at 2, 5, 7, 10, 15, and 30 days of age. Littermates were sham operated. Both groups were killed 72 hours later and the remaining adrenal removed and weighed. Control rats were killed at the time of the operation and the left adrenal gland was weighed. A consequence of the unilateral adrenalectomy is to cause the remaining gland to increase in size as a response to the relative increase in plasma ACTH per unit of adrenal tissue, and this increase can be quantified by determining the percent of hypertropy. Figure 9 shows the results of this study. Hypertrophy was high at 2 days of age, dropping to a minimum at 7, and then rising through 30 days of age.

The failure to get a good adrenal response in the 7-day group shown in Figure 9 could be due to the inability of the gland itself to react or it could be due to failure of the pituitary to secrete ACTH. Therefore, Zarrow, Philpott, and Denenberg (1968) carried out a second experiment in which they injected ACTH at 2, 7, and 21 days of age. Other animals received a beeswax-oil injection (this was the vehicle within which the ACTH was contained), and still other animals received no treatment. All animals were killed 48 hours after the injection, both adrenals were removed, and percent hypertrophy was calculated. Figure 10 shows the hypertrophy as a function of age and also as a function of the dose level of ACTH. At each dose level the 7-day group had a greater hypertrophy than either the 2-day or the 21-day group, thereby establishing that the adrenal gland of the 7-day old animal is actually more sensitive to the presence of ACTH than the gland of the 2- and 21-day old animals.

These results suggest the a 7-day old rat is deficient in its pituitary ACTH secretion. One method to test this hypothesis is to "prime" the animal's adrenals by a prior injection of ACTH, and then determine whether the primed gland is now more responsive to injections of ACTH. This experiment was carried out by Philpott, Zarrow, and Denenberg (1969) who injected experimental animals with two International Units of ACTH at 5 days of age while the nonprimed group of controls was given a beeswax-oil injection. Two days later

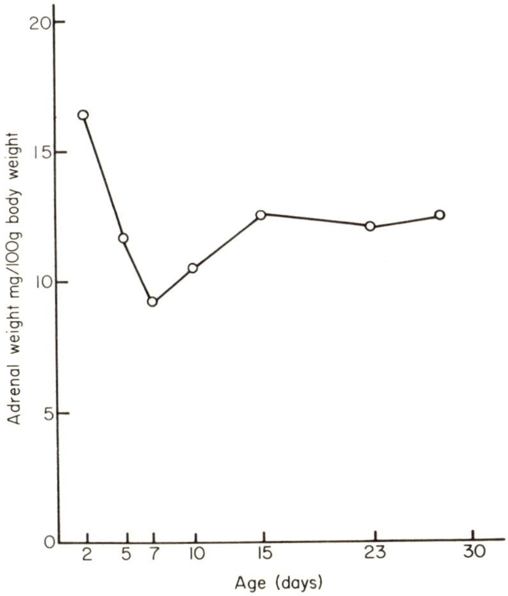


**Fig. 8.** Changes in the relative adrenal weight of the immature female rat between 2 and 30 days of age.

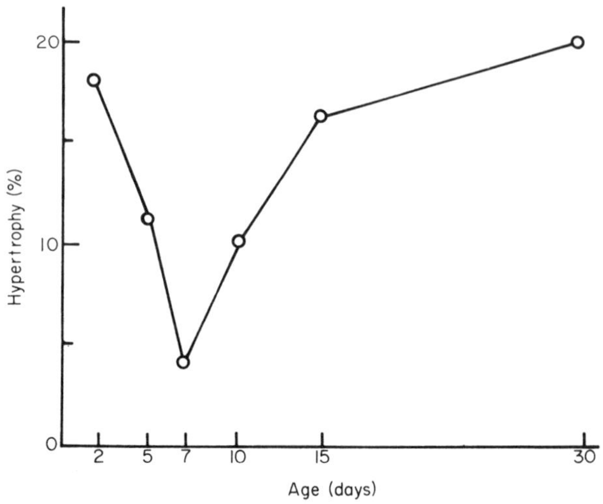


**Fig. 9.** Hypertrophy of the adrenal gland of the immature female rat following unilateral adrenalectomy. From Zarrow *et al.* (1968).

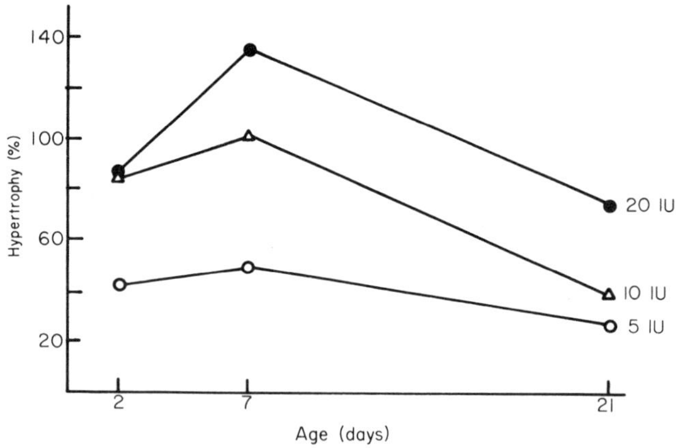


**Fig. 10.** Percent hypertrophy of the adrenal gland of the immature female rat as a function of age following treatment with ACTH suspended in beeswax oil. From Zarrow *et al.* (1968)

control animals from both groups were killed immediately to obtain a resting level while the remaining animals were given test doses of ACTH, and blood was collected 40 minutes later. Results of the experiment are shown in Figure 11.

The nonprimed 7-day old rat showed an increase in amount of corticosterone release as a function of ACTH injection, but this leveled off after 2 mU

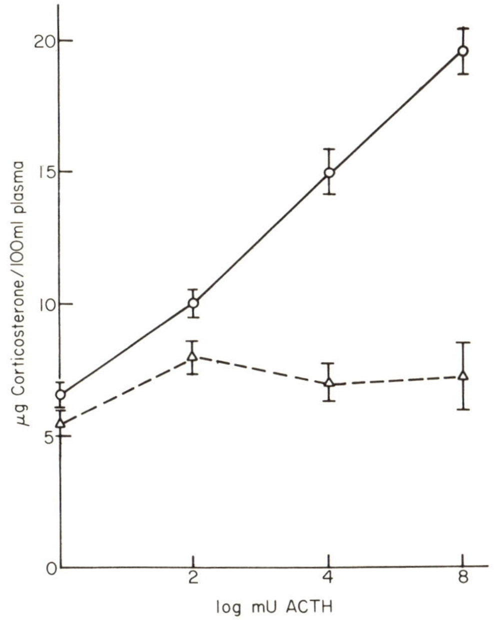


**Fig. 11.** Response of primed (O–O) vs. nonprimed (Δ---Δ) 7-days old rats to ACTH in 1% albumin. Rats were primed with 2 IU of ACTH in beeswax oil on day 5 of age, injected with test doses of ACTH in 1% albumin on day 7 of age, and blood was collected for corticosterone levels 40 min later. From Philpott *et al.* (1969).

of ACTH. On the other hand, those animals which had been primed at 5 days of age showed a linear response to the increasing amounts of ACTH injected.

These several results indicate that the biphasic pattern resulting from the lowered response of the 7-day old animal is probably due to the lack of circulating ACTH in the blood. Such a failure would bring about a lesser adrenal hypertrophy as well as a lesser corticosterone response to stressors. Since ACTH is present in adequate amounts at Day 2, the question arises as to why it is lacking in the 7-day old animal. No completely satisfactory answer is present at the moment. A possible explanation may be found in reports of Glydon (1957) and Campbell (1966) who report that the hypothalamic-portal system is not complete until 10-15 days of age. If this is true, then it is possible that the corticotropic releasing factor (CRF) can reach the pituitary gland of the late fetus or the early neonate by diffusion, but this would become more difficult as the animal grows. Eventually there would be a point reached, as seen in the 7-day old rat, when diffusion is inadequate and the hypothalamic-portal system has not yet been completed.

In summary, our studies investigating the physiology of the adrenal cortex and its relation to the pituitary gland indicates that there is a period of decreased responsivity of the gland at approximately 7 days of age. Further studies seem to indicate rather clearly that this decreased responsivity is not due to lack of the ability of the gland itself to respond, but is due, instead, to a decrease in the amount of circulating ACTH. The reason for the decrease in ACTH at this stage is not known, but may be related to the development of the hypothalamic-portal system.

## SUMMARY

This paper starts with a review of the data on the effects of handling as a technique for introducing stimulation in infancy. One of the major consequences of handling rats in early life is to bring about a reduction in emotional reactivity as measured by a variety of behavioral tests. We then discuss the effects of handling in infancy upon the corticosterone response of the adrenal cortex, and demonstrate an experimental relationship between behavioral measures of emotionality and the amount of corticosterone found in the adult animal.

Next, we briefly discuss some of the research involving neonatal sex hormones to develop the principle that the function of these hormones in very early life is to organize the brain with respect to maleness or femaleness while the function of the hormones in adulthood is to activate an already organized brain. Using this as a model we propose that stimulation in infancy acts to release corticosterone from the adrenal cortex, that this hormone acts upon the brain, presumably the hypothalamus, to modify neural organization and to make an animal less emotional. We then present our experimental data based on this hypothesis. To date all experiments we have carried out have been consistent with the hypothesis, but as yet the data are still not adequate to draw any firm conclusions concerning the validity of the hypothesis. Finally, we briefly discuss some of our research investigating the physiological characteristics of the adrenal cortex of the newborn animal and its relation to the pituitary gland.

## REFERENCES

Ader, R., & Belfer, M. S. Prenatal maternal anxiety and offspring emotionality in the rat. *Psychological Reports,* 1962, vol. 10, 711-718.

Ader, R., & Conklin, P. M. Handling of pregnant rats: Effects on emotionality of their offspring. *Science,* 1963, vol. 142, 411-412.

Campbell, H. J. The development of the primary portal plexus in the median eminance of the rabbit. *Journal of Anatomy,* 1966, vol. 100, 381-387.

Denenberg, V. H. Critical periods, stimulus input, and emotional reactivity: A theory of infantile stimulation. *Psychological Review,* 1964, vol. 71, 335-351.

Denenberg, V. H. Stimulation in infancy, emotional reactivity, and exploratory behavior. In D. C. Glass (Ed.), *Neurophysiology and emotion.* New York: Rockefeller University Press and Russell Sage Foundation, 1967.

Denenberg, V. H. The effects of early experience. In E. S. E. Hafez (Ed.), *The behaviour of domestic animals.* London: Bailliere, Tindall and Cassell, 1969. (a).

Denenberg, V. H. Open-field behavior in the rat: What does it mean? *Annals of the New York Academy of Science, 1969,* vol. 159, 852-859. (b).

Denenberg, V. H. The mother as a motivator. In W. J. Arnold (Ed.), *Nebraska symposium on motivation.* Lincoln, Neb.: University of Nebraska Press, 1970, in press.

Denenberg, V. H., Brumaghim, J. T., Haltmeyer, G. C., & Zarrow, M. X. Increased adreno-cortical activity in the neonatal rat following handling. *Endocrinology,* 1967, vol. 81, 1047-1052.

Denenberg, V. H., & Haltmeyer, G. C. Test of the monotonicity hypothesis concerning infantile stimulation and emotional reactivity. *Journal of Comparative and Physiological Psychology,* 1967, vol. 63, 394-396.

Fortier, C. Dual control of adrenocorticotropin release. *Endocrinology,* 1951, vol. 49, 782-788.

Fortier, C. Nervous control of ACTH secretion. In G. W. Harris and B. T. Donovan (Eds.), *The pituitary gland,* Vol. 2. Berkeley: University of California Press, 1966.

Glydon, R. St. J. The development of the blood supply of the pituitary in the albino rat, with special reference to the portal vessels, *Journal of Anatomy,* 1957, vol. 91, 237-244.

Haltmeyer, G. C., Denenberg, V. H., Thatcher, J., & Zarrow, M. X. Response of the adrenal cortex of the neonatal rat after subjection to stress. *Nature (London),* 1966, vol. 212, 1371-1373.

Haltmeyer, G. C., Denenberg, V. H., & Zarrow, M. X. Modification of the plasma cortico-sterone response as a function of infantile stimulation and electric shock parameters. *Physiology and Behavior,* 1967, vol. 2, 61-63.

Hebb, D. O. *The organization of behavior.* New York, New York: Wiley, 1949.

Hess, J. L., Denenberg, V. H., Zarrow, M. X., & Pfeifer, W. D. Modification of the corticosterone response curve as a function of handling in infancy. *Physiology and Behavior,* 1969, vol. 4, 109-111.

Hunt, H. F., & Otis, L. S. Restricted experience and "timidity" in the rat. *American Psychologist,* 1955, vol. 10, 432. (Abstract)

Hunt, H. F., & Otis, L. S. Early "experience" and its effects on later behavioral process in rats: I. Initial experiments. *Transactions of the New York Academy of Science,* 1963, vol. 25, 858-870.

Levine, S. A further study of infantile handling and adult avoidance learning. *Journal of Personality,* 1956, vol. 25, 70-80.

Levine, S. Plasma-free corticosteroid response to electric shock in rats stimulated in infancy. *Science,* 1962, vol. 135, 795-796. (a).

Levine, S. The psychophysiological effects of infantile stimulation. In E. L. Bliss (Ed.), *Roots of behavior.* New York: Harper, 1962. (b).

Levine, S. Sex differences in the brain. *Scientific American,* 1966, vol. 214, 84-90.

Levine, S., & Mullins, R. F., Jr. Hormonal influence on brain organization in infant rats. *Science,* 1966, vol. 152, 1585-1592.

Levine, S., Chevalier, J. A., & Korchin, S. J. The effects of early shock and handling on later avoidance learning. *Journal of Personality,* 1956, vol. 24, 475-493.

Levine, S., Haltmeyer, G. C., Karas, G. G., & Denenberg, V. H. Physiological and behavioral effects of infantile stimulation. *Physiology and Behavior,* 1967, vol. 2, 55-59.

McEwen, B. S., Weiss, J. M., & Schwartz, L. S. Selective retention of corticosterone by limbic structures in rat brain. *Nature (London),* 1968, vol. 220, 911-912.

Newton, G., & Levine, S. (Eds.) *Early experience and behavior.* Springfield, Ill.: Charles C. Thomas, 1968.

Philpott, J. E., Zarrow, M. X., & Denenberg, V. H. Prevention of drop in adrenocortical activity in the 7-day-old rat by pretreatment with ACTH. *Proceedings of the Society for Experimental Biology and Medicine,* 1969, vol. 131, 26-29.

Schapiro, S. Pituitary ACTH and compensatory adrenal hypertrophy in stress-nonresponsive infant rats. *Endocrinology,* 1962, vol. 71, 986-989.

Schapiro, S., Geller, E., & Eiduson, S. Neonatal adrenal cortical response to stress and vasopressin. *Proceedings of the Society for Experimental Biology and Medicine,* 1962, vol. 109, 937-941.

Thompson, W. R. Influence of prenatal maternal anxiety on emotionality in young rats. *Science,* 1957, vol. 125, 698-699.

Whimbey, A. E., & Denenberg, V. H. Experimental programming of life histories: the factor structure underlying experimentally created individual differences. *Behaviour,* 1967, vol. 29, 296-314. (a).

Whimbey, A. E., & Denenberg, V. H. Two independent behavioral dimensions in open-field performance, *Journal of Comparative and Physiological Psychology,* 1967, vol. 63, 500-504. (b).

Young, W. C., Goy, R. W., & Phoenix, C. H. Hormones and sexual behavior. *Science,* 1964, vol. 143, 212-218.

Zarrow, M. X., Denenberg, V. H., Haltmeyer, G. C., & Brumaghim, J. T. Plasma and adrenal corticosterone levels following exposure of the two-day-old rat to various stressors. *Proceedings of the Society for Experimental Biology and Medicine,* 1967, vol. 125, 113-116.

Zarrow, M. X., Haltmeyer, G. C., Denenberg, V. H., & Thatcher, J. Response of the infantile rat to stress. *Endocrinology,* 1966, vol. 79, 631-634.

Zarrow, M. X., Philpott, J. E., & Denenberg, V. H. Postnatal changes in the pituitary-adrenal axis of the rat. *Proceedings of the Society for Experimental Biology and Medicine,* 1968, vol. 128, 269-272.

Zarrow, M. X., Philpott, J. E., & Denenberg, V. H. Passage of $^{14}$C-4-corticosterone from the rat mother to the fetus and neonate. *Nature (London),* 1970, vol. 226, 1058-1059. (a).

Zarrow, M. X., Philpott, J. E., & Denenberg, V. H. Responsiveness of the adrenal gland of the neonatal rat. In S. Kazda and V. H. Denenberg (Eds.), *The Postnatal Development of Phenotype.* Prague: Academia, 1970. (b).

Zarrow, M. X., Philpott, J. E., Denenberg, V. H., & O'Connor, W. B. Localization of $^{14}$C-4-corticosterone in the 2-day old rat and a consideration of the mechanism involved in early handling. *Nature (London),* 1968, vol. 218, 1264-1265.

## DISCUSSION

Following Dr. Denenberg's presentation, a participant inquired about methodological problems of obtaining uncontaminated results in studies involving injection of corticosterone into rat pups. Is not the act of injection in itself a manipulation of the animal which can stimulate a corticosterone response? An experiment has recently been completed which helps to overcome this obstacle, Dr. Denenberg replied. Instead of injecting the rat pups, the investigator injects the pregnant mother, and the substance crosses the placental barrier. This is a new procedure, but it appears that it will give less ambiguous test results than direct injection of the pups.

How well developed is the human adrenal-cortical system at birth, a participant inquired. Can postnatal manipulations be done, or must this be prenatal? This is an area that has not been explored extensively, Dr. Denenberg replied, but it offers interesting possibilities for research. The few studies of newborn blood that have been done show high levels of corticosterone at birth, which drop off over the first 6 or 7 days of life.

In prenatal rats, a participant said, adrenal function is present after the 18th day of pregnancy and the adrenal is very large; but as the fetus develops, the adrenal ceases to grow. This may have something to do with the adrenal lag that occurs shortly after birth.

Dr. Denenberg has indicated that he measures the corticosterone in the plasma. Perhaps more information could be obtained about the onset and duration of response by histochemical measurements. Investigators at Baylor University Medical School have found a releasing factor in the hypothalamus with corticotropic hormone. It is possible the hypothalamus may provide a releasing hormone for the corticosterone. Studies on the corticotropic releasing factor are also under way in his laboratory, Dr. Denenberg said.

Participants discussed several other methodological problems. Incomplete reporting often makes it difficult to compare or evaluate studies, they said. For example, many animal studies fail to report such important variables as the strain of rats used, their weight, whether the animals were born in the laboratory or imported, where the mother was born, what types of cage and litter material were used, what diet was used, or what the handling and cleaning schedule was. Other possible influences such as the effects of odor residues left on the stimulus objects by previous litters need to be considered.

Litter size and the methods used to reduce the litters are other important variables which should be clearly identified in reports of studies. Failure to keep litter size constant builds in a source of either contamination or error. The physical manipulations of the experimenter have a direct effect on the animal. So does "handling" by the mother rat. She must be considered a mediator in some of the events, and how much she stimulates each pup depends in part on the size of the litter. With small litters she spends more time manipulating each pup.

Summarizing the methodological problems, Dr. Denenberg said that in development, everything is important. The experimenter is working with a set of confounding variables, some of which can be unconfounded by experimental procedures; but he has to choose what he wants to work with and which ones he can afford to ignore without prejudicing the results.

A participant asked for clarification of Dr. Denenberg's comment about rats organizing into a male or female brain. Up to a certain point, Dr. Denenberg replied, it is possible for the brain to develop in either direction, toward a male orientation or toward a female orientation. Experimental procedures have shown that it is possible to make a genetically female animal behave in masculine fashion, and vice versa, by manipulating the physiological environment. This is one of the best examples of organization that we have.

In reply to another question, Dr. Denenberg said that physiological effects of stress appear earlier than cognitive effects both in animals and in humans. A newborn animal can learn, but cannot retain information as well as an older animal. Physiological responses indicative of emotionality can be produced in rats at the age of 5 to 10 days; but if the goal is to produce better cognitive, perceptual, problem-solving behavior, it is necessary to wait until the animal is 18 days old. The rat does not show retention before this age, perhaps because the neural system is not sufficiently developed for long-term retention.

This is true of avoidance condition in animals, a participant agreed, but cognitive effects can be observed earlier if the experiment relates to food-getting behavior, which is the animal's principal activity at that stage of his life. Very young animals discriminate odors and tactile form when these are related to nursing.

Subtle precursors of cognitive effects can be observed in rats introduced to unfamiliar objects, another participant said. When exposed to unfamiliar objects, animals raised in an enriched environment sniff them and move around them in a sort of cautious exploration. Animals raised in a deprived environment are more apt to crash into the objects and be startled by them. However, in early experience work it is very difficult to find dependent measures that are satisfactory in assessing conceptual variables. In the Hebb–Williams maze, for example, the animals' ability to find their way through it is confounded by their tendency to explore the maze and by their sheer motoric activity.

Handling in early infancy is one factor that affects stimulus-seeking behavior, Dr. Denenberg said. An animal that has been handled shows more interest in manipulating novel objects. Emotionality is orthogonal to exploratory behavior, and it is possible to affect emotionality independently of exploration. However, these things do not affect the way the animal performs in the Hebb–Williams maze. It is true that the maze is not a very satisfactory apparatus, but at present there is nothing better to use.

In his experiments, Dr. Denenberg said, litters were reduced to eight, with a high preponderance of females. Some litters were handled, others were not. Both groups were bred. The rat pups were cross-fostered so that handled mothers raised either offspring of handled mothers or offspring of nonhandled mothers. The rat pups were not handled except at weaning; thus, any effects obtained had to be mediated through the mother, which means the effects had to be functions of what happened to the mother when she herself was an infant. The body weight of the rat pups was measured at 21 days and their open-field activity at 50 days. There were weight differences between pups of handled and non-handled mothers at weaning, but not in adulthood. However, a significant interaction with respect to open-field activity in adulthood was identified. Offspring of nonhandled mothers reared by handled mothers behaved like nonhandled females. This represents a nongenetic effect in the young rats.

A second study was done which examined the preweaning and postweaning activities of rat pups whose grandmothers had been either handled or not handled. The data show that behavior of the young rat is determined in part by what happened to the animal's grandmother when she was an infant in combination with what happened to its mother during her early life.

Dr. Denenberg said that it is his strong belief that handling does have biological meaning with respect to the animal and that the analog of this relates to maternal manipulation. The handling procedure brings about a slight temperature loss, but the temperature loss alone does not explain the later effects. After handling, the animals are put back with their mothers, who engaged in such behavior as picking up the pups and running around with them. It may be that these maternal manipulations are a major factor in the long term effects that have been observed.

Some experiments done as long ago as the 1920s with mollusks and fly grubs showed that certain adaptations of the organism to the environmental input were not reversed when the organism was returned to its normal environment, a participant said. This may have a bearing on the effects Dr. Denenberg is observing in the progeny of mothers handled during their own infancy.

Some of Dr. Denenberg's comments about nongenetic two-generational changes seem to tie in with Valadian's recent nutritional studies, a participant said. Using growth data gathered at Harvard during the 1920s and 1930s, Dr. Valadian has shown that the nature and character of offspring born to young women who were the subjects of the earlier studies are clearly related to the mothers' nutritional status at puberty, long before they began to reproduce. Dr. Valadian's studies have centered on human nutritional status, while Dr. Denenberg's seem to indicate that what happens to the mother rat prior to conception can affect the behavior of her offspring. The interlinking of the findings of these two studies could point directions for further exploration of how various qualities of potential offspring can be shaped by preconceptual influences on the mother.

Do any of Dr. Denenberg's data support the concept expressed many years ago that certain types of performance are produced under conditions of optimal stress, participants inquired. How is optimal level of stress determined? Is the optimal level of stress whatever level exists at the time the animal performs a target behavior? Yes, Dr. Denenberg replied. It is possible to manipulate conditions to optimize performance on a particular criterion of measurement. However, this may injure other performances; optimal stress is not the same for all types of behavior.

There is much evidence in the literature that stress influences progress in many different kinds of learning situations, a participant said. Sometimes this facilitates learning; sometimes it impairs it. That is why it is difficult to accept the distinction made by some participants between cognitive and emotional activities in the very young organism. What is needed are studies to determine what alterations in the states of infants produce what kinds of differential effects in learning.

The whole question of the effects of stress during early life needs further exploration, several of the participants said. Studies have attempted to relate early stress to adult stature, age of onset of menarche, intelligence, and other variables. Stress has been variously defined as cultural molding, severe illness, or such events as smallpox inoculation. Some related studies have been done in animals; for example, one study done in Dr. Denenberg's laboratory has shown that both male and female rats that are handled in infancy are sexually precocious. At the present time, however, such studies in humans tend to be frustrating and inconclusive because of the difficulty of controlling the many variables.

There is also a great need for studies to determine whether the biochemical and behavioral data obtained separately by Dr. Denenberg and Dr. Rosenzweig are independent of each other or related. Such studies should be helpful in resolving some apparent contradictions. Dr. Denenberg's work, for example, suggests a critical period early in life; Dr. Rosenzweig's does not. It is obviously impossible for these investigators to duplicate each other's laboratories, and many methodological problems would have to be solved, but it should be possible to develop some joint studies.

## SPECIAL REMARKS OF DR. LEWIS LIPSITT

Supplementing the reports of Dr. Rosenzweig and Dr. Denenberg on the effects of different types of experience on the development of regulatory mechanisms in animals, Dr. Lewis P. Lipsitt reported briefly on studies of early conditioning in human infants done in his own laboratory and elsewhere. Dr. Lipsitt is Professor of Psychology and Director of the Child Study Center at Brown University. Studies have shown that newborn infants cannot only detect odors but can discriminate between them, Dr. Lipsitt said. They can habituate to certain odors systematically presented to them and then, when a distinctive or novel odorant is administered, react in a way that indicates they are aware of a difference.

Considerable evidence is also available to show that newborn infants respond to both classical and operant conditioning procedures. If one reinforces the infant for engaging in such operant behavior as head turning or sucking, the behavior becomes enhanced as a function of the learning circumstances administered to the child.

Evidence is available which shows that the human newborn is capable of cognitive functioning; he will acquire new responses or demonstrate potentiation of responses as a function of conditioning and reinforcement circumstances. Even premature infants are conditionable as much as a month before the time they would have been born at full term. They will, for example, learn to keep their eyes open if a bottle is administered whenever they open their eyes. Similarly, they will alter their vocal and crying behavior as a function of certain stimulus conditions.

The premature child is an organism which is in developmental jeopardy in a number of different respects from birth onward. Premature children contribute heavily to the populations of nonreaders, to the populations that require treatment in child guidance clinics, and to the populations of school deficit children. Until recently, little has been known about what kinds of trouble the premature child has in growing up. It has been widely thought that the premature child is in some respects a defective organism at birth, that this is why he was premature to begin with, and that he simply suffers for the next few years from the deficit with which he was born.

An alternative hypothesis holds that the deficits are instilled in the premature child as a result of environmental deprivations. In a typical good premature nursery, the infant exists in an environment largely devoid of variable stimulation. He lies in an incubator which has a motor which often roars at high intensity. He is surrounded by white and sees very little color. The plastic dome over his head diffuses the light and makes it difficult to see beyond the dome with any clarity. Thus, during the first crucial weeks or months of his life, the child is subjected to a deprived environment.

In the premature nursery at Providence Lying-In Hospital, the babies have been separated into two groups by Dr. E. R. Siqueland, an associate of Dr. Lipsitt. The infants in the control group receive the standard premature care. They are picked up to have their diapers changed, and they are taken out of the incubators to be fed occasionally, if their condition permits. Infants in the experimental group are picked up by the nurses, rocked, sung to, patted or stroked, and generally given all the tender loving care normal, full term babies receive. Duration of this exposure varies, as prematures remain in the hospital for different lengths of time, and no baby is kept in the hospital longer than necessary.

In a later part of the study, the babies are tested in an operant conditioning situation. This involves presenting a nipple to the infant. When he sucks on it, he illuminates a screen that he can see in front of him. Followup studies when the babies are 3 months old show that the infants who have received this extra stimulation are better learners than those who have not. Several sets of twins have been included in the study, with one twin in the experimental group during the premature period and the other in the nonstimulated control group. The siblings show distinct differences in learning ability.

One can get a striking amount of conditioned operant head-turning even in 2- and 3-day old babies, Dr. Lipsitt said. When the rooting reflex is used to get the baby to turn his head in the direction of the side of his mouth that has been touched, and he is rewarded by being fed, the incidence level of head turning rises from 25% to 80% within a half hour conditioning period.

Within a half hour it is also possible to condition the infant to turn his head only to one side, as shown in a study by Drs. Siqueland and Lipsitt. If tone A, a gong-like sound, is paired with touch to the right cheek, and if head-turns to the tone A plus touch stimulation is rewarded with feeding, while the presentation of tone B, a buzzer, is also paired with touch to the right cheek but does *not* result in reinforcement for head-turning, the newborn infant will learn to respond differently to the two tone–touch combinations. Moreover, if the stimulus combinations are then reversed, so that tone B is reinforced while tone A is not, a reversal of conditioning can be obtained within a half hour. The child not only learns a correct response under rather complex stimulating conditions, but he can unlearn one response and learn the opposite.

In another study, the head of each newborn infant was placed in a harness attached to a protractor in such a way that the experimenter could see the number of degrees the child turned his head. Whenever the infant turned his head a certain number of degrees, the bottle was given. Infants treated this way sharply increased their head turning behavior as compared to a control group that received the same amount of food without the reward being contingent upon head turning.

Early conditioning is important because the child has an opportunity to learn that what he does makes a difference. This is a lesson all human beings have to learn eventually. Although the issue remains speculative, it seems possible that there are critical periods in development with regard to learning certain contingencies. That is, the child may have to learn within a certain age range that what he does with a nipple makes a difference whether or not he gets milk from it. If he does not learn this by a certain age, he may become deficient with respect to that particular response. Moreover, most of what we do in life has to do with getting sensory and perceptual feedback in response to our behavior. For example, a child sucking on a nipple gets milk commensurate in amount with the intensity of his sucking response. In the normal course of development, he subjects himself to a lot of learning in this respect. He learns that if he sucks too hard, he gets too much and may gag; and if he does not suck intensively enough, he may not get enough milk. So he learns to adjust his behavior just as an adult learns to adjust the spigot of a water fountain, or the accelerator of a car to produce the desired result. Many of the social behaviors children learn during the first few months of life are of this sort; for example, a child quickly learns that he can elicit a smile from his mother by smiling at her. Children learn to behave in certain ways in order to receive reinforcement and stimulation of certain sorts that they find desirable.

Twenty-five or 30 years ago many prematures were unable to survive because of the limitations of the medical care then available, a physician said. The infants were often kept in what was known as a Hess bed. This was merely a water-jacketed bassinet with a cover over the top part to prevent air conduction currents from carrying away body warmth, and the warmth supplied by the water jacket. However, these babies could look out and see a variety of sights and activities within the traditional hospital room. It would be interesting to compare the development of surviving prematures of that era with the development of infants isolated in the plasticized, motorized units which have been considered ideal environmental chambers for them in recent years, but unfortunately, few critical evaluative studies were done in those days.

Dr. Clifford Barnett and a group of investigators at Stanford have been studying the effects of different types of premature care on the relationship between the mother and the infant, and on the child's subsequent development, another participant said. They feel that prematures are often deprived of stimulation not only in the hospital, but after they go home, because the separation of mother and infant during the early weeks of life interrupts establishment of the mother's attachment for her child. When the child goes home, the mother fails to stimulate it with the normal amount of mothering behavior. The group has found that the quality of mothering is quite different when the baby goes home if the mother is encouraged to come to the nursery each day to hold and feed and talk to her baby. This is true even of babies who must remain in isolettes, but whose mothers put on masks and gloves to caress the infants without picking them up.

# DETERMINANTS OF AN INFANT'S CARDIAC RESPONSE TO STIMULATION[1]

*Alfred Steinschneider*

State University of New York (Syracuse)

During the past decade or so, behavioral scientists have become increasingly involved in the elaboration of autonomic nervous system responses associated with a variety of psychological processes. A reflection of this increased awareness of the potential importance of autonomic activity toward an understanding of behavior is the relatively recent formation of the Society for Psychophysiological Research and the establishment of its journal, *Psychophysiology: The Journal of Objective Research in the Physiology of Behavior.*

Within this same decade students of child development have also become interested in the study of autonomic reactions. Some of this increased theoretical and research activity stems from the recognition that the neonate is not a passive recipient of environmental inputs. Rather he responds to environmental demands in an individually characteristic manner which has the potential for modifying the interactional process. Furthermore, it had been proposed that an elaboration of these individually characteristic response patterns might lead to a greater understanding of the development of personality, emotions, attention, and learning.

[1]This work was supported in part by a grant from the National Institutes of Health, 2RO1MHO4605, and an institutional grant from the Upstate Medical Center, 11-8505.

Although considerable technical advances have been made, relatively few autonomic response systems have been studied in infancy. Changes in skin resistance, a response system frequently employed in adult research, has not been fruitfully explored in early infancy. This stems in part from the observation that stimuli do not produce many skin resistance responses in the neonate (Crowell, Davis, Chun, & Spellacy, 1965). Furthermore, studies have not employed vasomotor responses because of the frequent interference of the response by movement artifacts.

Cardiac rate has been one of the most commonly used autonomic response systems in the study of infants. Technically this is one of the easiest response measures to obtain. A further stimulant to the study of cardiac rate responses derived from the hypothesis that the *direction* of cardiac rate change to stimulation reflects a changing sensitivity to external stimuli. Thus Lacey and his colleagues (Lacey, 1959; Lacey, Kagan, Lacey, & Moss, 1963) have proposed that cardiac rate deceleration is associated with efforts to increase "environmental intake," whereas an increase in cardiac rate is a correlate of efforts to decrease the effectiveness of external stimuli. In keeping with this formulation, Graham and Clifton (1966) have argued that cardiac rate deceleration is a component of the orientation reflex and cardiac rate acceleration a component of the defense reflex.

Fundamental to the study of these formulations are those efforts designed to define the directional aspects of the cardiac response to stimulation. Examples of questions posed by many of these studies include: Is the response to a nonsignal stimulus monophasic, biphasic, or triphasic? If it is monophasic, is it basically a decelerative or an accelerative response? Similar interests have also led to the study of the role of the behavioral state. For example, it has been suggested that the orientation reflex is most apt to occur in the waking state. Furthermore, the defense reaction, a set of responses which are associated with efforts to limit the effect of subsequent stimulation, is most apt to occur during sleep.

On the basis of a number of studies performed in our own laboratory as well as those of others, it is apparent that the responsiveness of neonates and children within the first few years of life is not solely dependent upon the characteristic of the stimulating conditions. This does not deny the importance of the stimulus. In fact, in the course of this report we will make note of the effect of stimulus variables on the response. However, we would like to emphasize that to a very large extent the magnitude as well as the directional aspects of the response are dependent upon organismic variables preceding the onset of stimulation. Very detailed examination of the cardiac rate response to a standard stimulus has revealed that the very same neonate will demonstrate a considerable amount of variation in the *form* of the cardiac rate response.

In this report we will present some of the data supporting the conclusion that

1. Stimulus properties affect the magnitude and direction of the cardiac rate response in infancy.
2. The magnitude as well as the direction of the cardiac response are determined to a large extent by the level of organ function prior to stimulation.
3. There is a considerable degree of individual differences in responsiveness.
4. Developmental factors influence the form of the cardiac response to stimulation.
5. The effect of behavioral state on the cardiac response is developmentally related.

## GENERAL PROCEDURE

Data obtained primarily from six different studies will be considered. Reports on two of these experiments have been published previously (Lipton, Steinschneider, & Richmond, 1966; Steinschneider, 1968; Steinschneider, Lipton, & Richmond, 1966). Three of these experiments employed normal newborn infants (age 2–5 days). In two of the studies the subjects were between the ages of 1 and 4 years. In the remaining study a group of subjects were studied as newborns and again at 2½ and 5 months of age. A brief description of the various studies from which we will draw is presented in Table 1.

TABLE 1
*Summary Description of Studies*

| Study | No. | Age | Stimulus |
|---|---|---|---|
| Intensity-Newborn[a] | 9 | 2–5 days | White noise: 55, 70, 85, 100 dB |
| Duration-Newborn | 8 | 2–5 days | Air stream: 1, 2, 5 sec |
| Intensity-Temperature-Newborn | 24 | 2–5 days | White noise: 75, 90 dB<br>Ambient temp: 75°, 90°F |
| Longitudinal[b] | 14 | 2–5 days, 2½ and 5 mos | Air stream |
| Infant | 10 | 1–4 years | White noise: 85 dB |
| Intensity-Infant | 12 | 1–4 years | White noise: 55, 70, 85 dB |

[a]Steinschneider *et al.* (1966); Steinschneider (1968).
[b]Lipton *et al.* (1966).

The auditory stimulus was produced by a General Radio White Noise Generator and covered the entire audible spectrum. It was fed through an Acrosound amplifier into an 8-inch Utah speaker housed in a cabinet near the subject. The intensity of the stimulus was measured at the subject's ear by means of a General Radio Sound Level Meter (Scale A). The duration of sound stimulation was maintained constant at 5 seconds.

The "air stream" stimulus consisted of a jet of air directed to an area above the level of the umbilicus. It was of sufficient intensity to produce minimal skin indentation.

In general, the same testing procedure was employed in all of these studies. The subject was brought into an environmentally controlled test room and efforts made to maintain him in a relaxed or sleeping state. Each subject received between 45 and 60 stimulus trials during a test session. The stimulus was presented when the heart rate was relatively stable, the subject was quiet and no less than 40 seconds had elapsed from the presentation of the previous stimulus.

## RESPONSE MEASURES

An example of the cardiac response to a single presentation of an air stream stimulus is presented in Figure 1. The lowest channel in this figure contains the electrocardiographic tracing. Visual examination of the electrocardiogram does not demonstrate readily the effect of stimulation on cardiac rate. However, this process is simplified by means of the cardiotachometer. The cardiotachometer measures the time interval between successive *r*-waves of the electrocardiogram, converts this interval to rate and plots this rate onto *x-y* coordinates, where the abscissa is time and ordinate is in beats per minute. The channel above the ECG trace contains the appropriate cardiotachometer transformation. It is now readily apparent that shortly following the onset of stimulation cardiac rate increased, reached a peak heart rate level, and then returned to near prestimulation levels.

A significant problem faced by the investigator employing cardiac rate is the choice of measures which would describe at least some aspect or aspects of this response curve. This choice is a crucial one, for to a large extent it imposes limits upon any conclusions that can be reached.

A number of different approaches have been employed, each of which has its own difficulties and limitations. Bridger and Reiser (1959) calculated the average heart rate during (*a*) the 5-second period immediately prior to stimulation, (*b*) the 5-second period during which the stimulus was presented, and (*c*) the 5-second period immediately following the stimulus presentation. They then computed the difference in rate between the average prestimulus level and either the average stimulus or poststimulus level, depending upon which level resulted

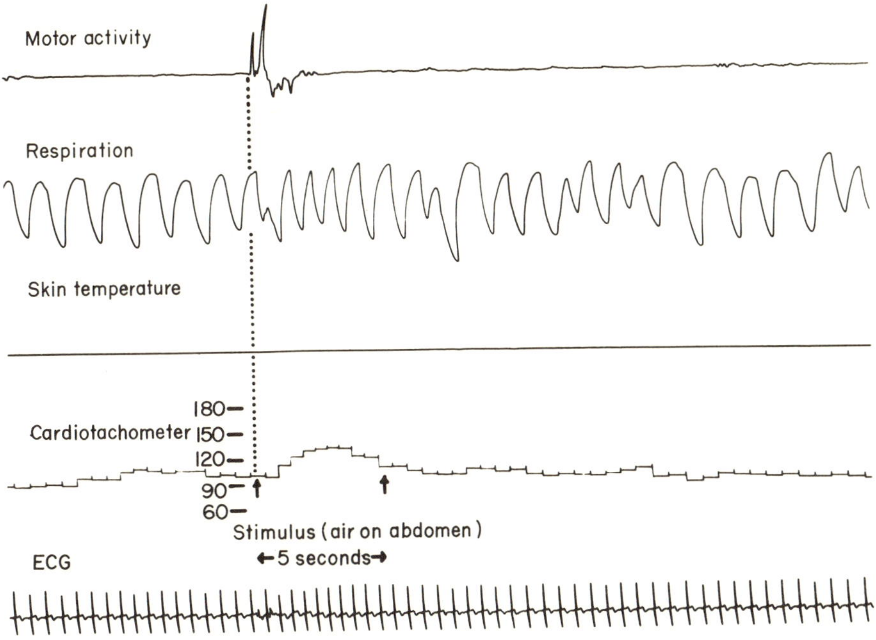


**Fig. 1.** Polygraph tracing indicating stimulus presentation, motor activity, respiration, skin temperature, cardiotachometer, and ECG responses. (From Lipton & Steinschneider, 1964.)

in a maximum difference consistent with the observed motor response. Thus if there was an increase in motor activity, that difference was selected which resulted in the greatest increase in heart rate, if motor activity decreased, that difference was chosen which yielded the greatest decrease in rate. This procedure suffers from a number of weaknesses. In the first place, it obligates a positive correlation between the motor response and the cardiac reaction. In addition, it completely negates the temporal characteristics of the cardiac response and results in an attenuated estimate of the maximal change induced by stimulation.

Another approach is one which attempts to maintain the total character of the response curve by obtaining the instantaneous heart rate at fixed time intervals and obtaining an average response curve by computing the mean at each of these time periods. Figure 2 is an example of the use of this method. It also points out one of its weaknesses. Let us assume that we have obtained two stimulus trials from the same individual. Furthermore, the cardiac response curves are identical except for its temporal characteristics. In essence, I am posing the situation in which identical maximum heart rates are obtained on both trials. The dashed line represents the curve determined by averaging the rates at fixed times. It can be seen from an examination of the average curve that it tends to underestimate the maximal increase in heart rate achieved in the two individual trial curves. Furthermore, the greater the amount of variability in the temporal aspects of the response curve, the greater will be the underestimation of the maximal response. Since we can anticipate that individuals as well as experimental conditions might have different temporal *variabilities* this approach is not adequate when concern is with estimating the maximal change in rate resulting from stimulation. A variant of this approach is to obtain an average across fixed *r–r* intervals, i.e., every third heart beat following stimulation. Unfortunately, this procedure does not resolve the difficulties already mentioned.

The procedure we are employing differs from those discussed and was developed following a preliminary, though detailed, inspection of the cardiotachometer tracings obtained from a number of different neonates. In a methodological paper published in 1961, we reported that . . .

Inspection of the individual trials reveals that, in general, stimulation of the newborn (e.g., with a 5-sec. air stream on the abdominal wall) produces a steady increase in heart rate that reaches a peak or maximum some seconds after the onset of stimulation and then decreases to a level approaching or less than that prior to stimulation. Furthermore, the rise to the peak as well as the return from the maximal heart rate often assumes the shape of an ogive.

There are, however, individual time curves (trials) that do not follow this general scheme. These assume one of two forms: In some the curve is the mirror image of that

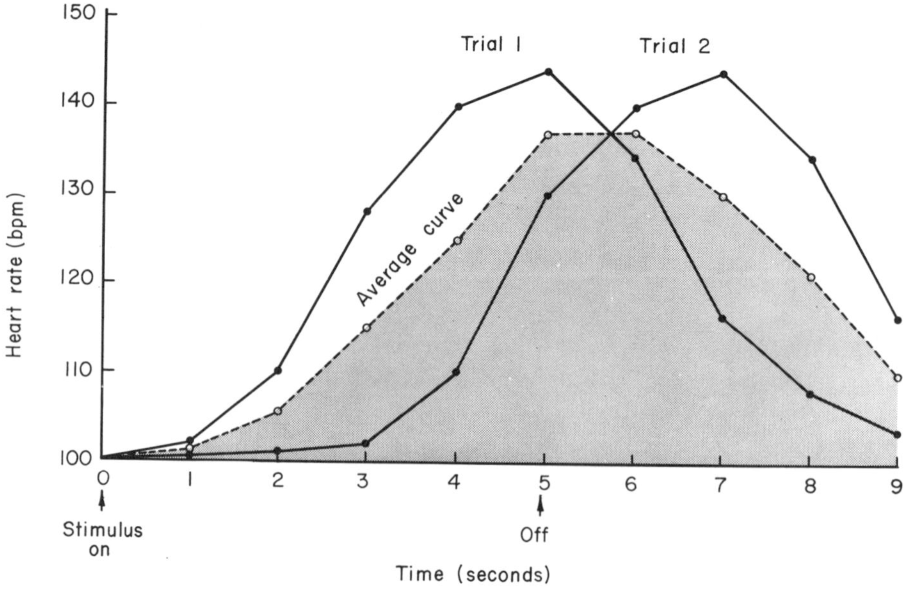


**Fig. 2.** The average response curve derived from a consideration of the individual response curves at fixed time periods. (From Lipton *et al.*, 1961.)

already described. In the remainder, there is no "response" to stimulation. That is, there is no change in heart rate following the onset of stimulation.

We can then place the individual response to stimulation on a continuum whose limits represent, on the one hand, a primary increase in heart rate and, on the other, a primary decrease with a "no response" being represented between the two [Lipton, Steinschneider, & Richmond, 1961, p. 465].

The next step was to define a set of response curve measures which, when recombined, would reasonably describe the total obtained cardiac rate response to a single stimulus or trial. Fundamental to this approach is the determination, for each response trial, of the initial direction of the cardiac rate response. This is accomplished by determining the directional change from the prestimulus level of the first six or more successive cardiac cycles which deviate in the same direction from the prestimulus heart rate. Once having defined the initial direction of the cardiac response, we then extract the various response curve measures.

The response curve measures we extract are schematically presented in Figure 3 and include prestimulus level– the instantaneous heart rate immediately prior to stimulus onset; peak level–the maximal heart rate level achieved (if the trend is one of acceleration) or the lowest heart rate level (if the heart rate decelerates); peak magnitude–the difference between the peak and prestimulus level; primary slope magnitude–the maximum difference in rate between two successive beats and before the attainment of the peak level; return level–the heart rate level of the response at which the response appears ended and the rate has returned toward or beyond the prestimulus level; secondary slope magnitude–the maximum difference in the rate between two successive beats following the peak and prior to the return. The temporal aspects of each response curve are also measured: primary slope time (*A*)– time interval between the onset of stimulation and the primary magnitude; peak-primary time (*B*)–time interval between the primary slope and peak level; secondary slope time (*C*)–interval between the peak and onset of the secondary slope; return-secondary time (*D*)–time interval from the onset of the secondary slope to the return level; peak time (*A+B*)–time interval from stimulus on to the occurrence of the peak level; return time (*C+D*)–time interval from peak level to the occurrence of the return level. Two additional measures should be mentioned: Base I magnitude is the change in the heart rate between the prestimulus level and the heart rate level from which the primary slope develops; prestimulus departure is the difference in rate between the return and prestimulus level. It should be emphasized that all of these measures are extracted from each stimulus response curve.

Figure 4 is a reconstruction of the response curve using the response measures extracted. It can be seen that the reconstructed curve does represent adequately the actual response.

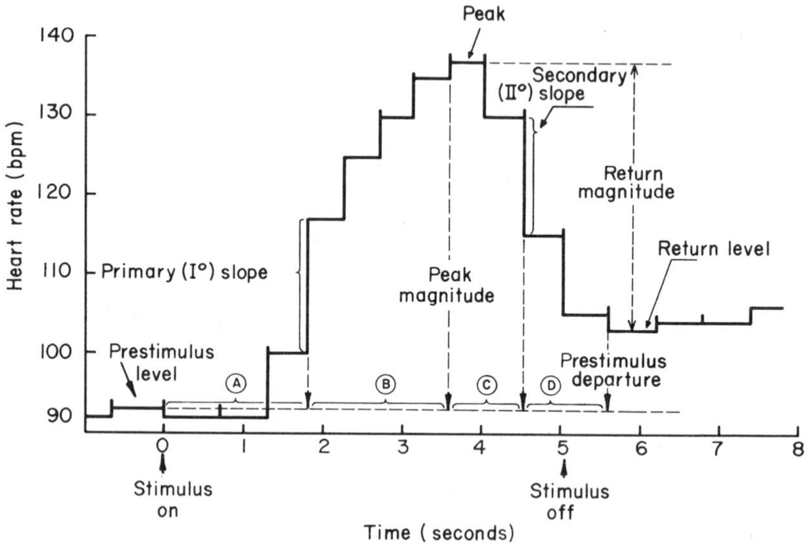


**Fig. 3.** Schematic representation of the cardiotachometer response to a single stimulus illustrating the various response curve measures extracted. (From Lipton *et al.*, 1966.)

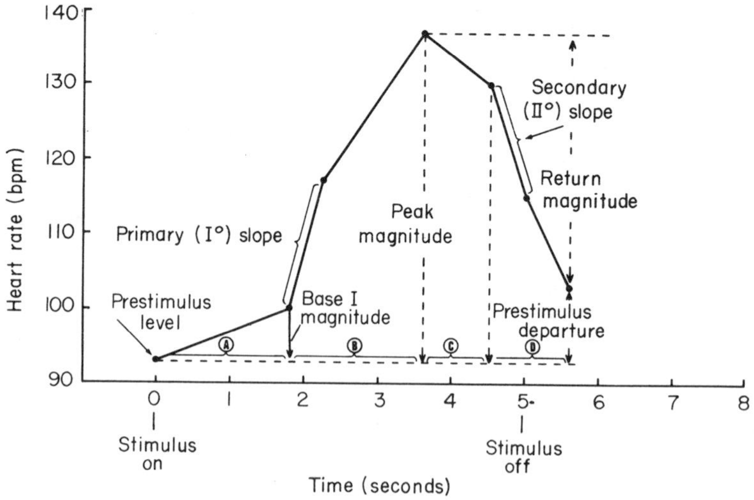


**Fig. 4.** Response curve reconstructed from measures extracted in Fig. 3. (From Lipton *et al.*, 1966.)

This approach has been subject to several criticisms (Graham & Jackson, 1970). One concerns the difficulty in defining the response measures in a manner that would allow for their extraction in a reliable and objective manner. Early in our research effort these response curve measures were extracted visually and one could argue that to some extent this approach was confounded by an element of subjectivity. However, following considerable efforts we were able to define these measures with sufficient objectivity to allow for the development of a computer program which now automatically extracts these measures from digital tape. A second criticism, and one which is perfectly valid, points to the rigidity of this approach; that is, the measures extracted, as well as the criteria for their determination, apply only to the family of response curves for which it was developed. However, adequate recognition of this limitation also points to what I believe to be an asset of this method. It obligates the investigator to inquire continually into the nature of the response curve being measured. This requirement should apply to any procedure employed by the researcher when attempting to quantify a complex response pattern.

It should also be made clear that there is nothing inherent in this approach that requires the measurement of all the measures described. One can, depending upon the needs of a specific study, select a given measure or measures for detailed evaluation. Thus the response measure used by McDonald, Johnson, and Hord (1964) is similar to our return magnitude whereas Roessler, Greenfield, and Alexander (1964), in a report published in the very same issue of *Psychophysiology,* selected a response measure similar to the peak magnitude.

## AVERAGE RESPONSE

By means of this analytic approach we have been able to demonstrate that the cardiac rate response is influenced by a number of variables.

A summary of the effect of sound intensity on the cardiac rate response is presented in Figure 5. This figure contains the average reconstructed cardiac rate response curves obtained from a single newborn exposed to four intensities of white noise. Four intensities were studied: 55, 70, 85, and 100 dB. By analyzing each response curve measure separately in a group of neonates, we noted that both the temporal as well as the magnitude aspects of the response curve systematically were modified by variations in sound intensity. Increasing the intensity of stimulation resulted in an increase in the primary slope, peak, secondary slope, and return magnitudes. In addition, the more intense sounds produced a decrease in the primary slope time and an increase in the peak minus primary slope time. Neither of the comparable return time measures was statistically altered by changes in sound intensity.

In this same study we investigated the extent to which individual neonates maintained their same degree of relative reactivity. This was done by correlating subject response measures across pairs of stimulus intensities. This analysis

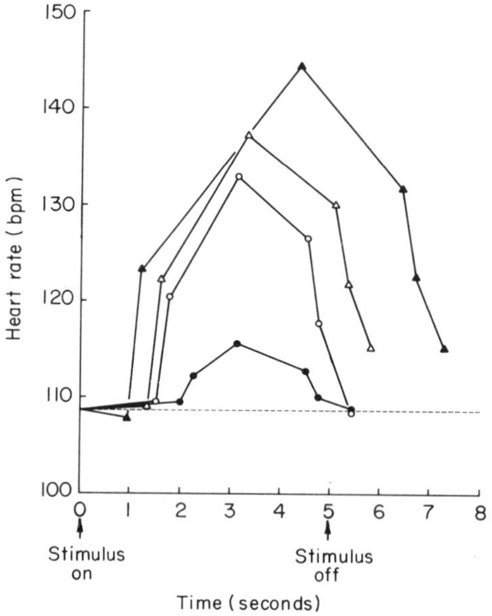


**Fig. 5.** Average reconstructed cardiac rate response curves derived from newborn Re. ● 55 dB; ○ 70 dB; △ 80 dB; ▲ 100 dB. (From Steinschneider *et al.*, 1966.)

revealed not only that neonates were reliably different from one another but that they also maintained the same degree of responsiveness relative to their peers although intensity was varied. Thus, for example, when considering maximal reactivity as the *subject* response measure, correlation coefficients for peak magnitude ranged between 0.72 and 0.93. Similar correlation coefficients for return magnitude varied between 0.52 and 0.92. Additional support for the concept of individual differences and stability of relative responsiveness was obtained when we examined the average primary slope and peak minus primary slope times. In essence, neonates with relatively large increases in heart rate to weak sounds also demonstrated relatively large increases in heart rate to loud sounds. Furthermore, individuals who had relatively short latencies (i.e., primary slope time) at one sound intensity responded rapidly when other sound levels were employed.

Similarly, stimulus intensity was found to influence the cardiac response in children between the ages of 1 and 4 years (Figure 6). These curves were reconstructed from the averages of the various response curve measures. Consistent with the newborn data, an increase in intensity resulted in an increase in the magnitude measures, a decrease in response latency (primary slope time) and an increase in the duration of the peak-primary and secondary slope times. Two interesting differences should be pointed out between these curves and those obtained on newborns. First, the older children demonstrated an initial decrease in heart rate prior to the primary slope magnitude and second, the return portion of the response brought the heart rate level below that of the prestimulus level.

Further support for the differences observed between the neonates and the older children was obtained from a group of subjects who were studied as newborns and again at 2½ and 5 months. The same stimulus, a 5-second duration air stream was employed at each of these ages.

The curves presented in Figure 7 were reconstructed from the average response measures obtained from a single subject and are similar to the curves obtained from the group as a whole. Examination of this figure reveals a number of age-related modifications of the response curve. In the newborn period the average cardiac response initially accelerated and was followed by a return to a heart level *above* the prestimulation level. By 2½ and 5 months of age the average response revealed an initial brief deceleration followed by an accelerative phase and a return to a level *below* that of the prestimulation heart rate. Upon closer inspection we can also see that by 2½ months there were smaller primary

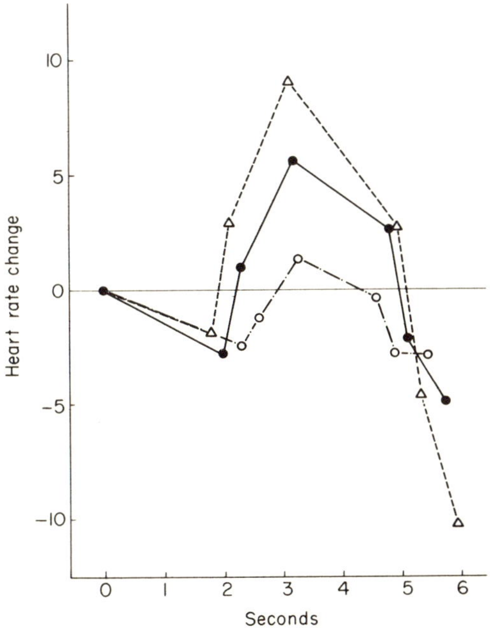


**Fig. 6.** Group average cardiac rate response curves at each of three stimulus intensities. Subjects were between 1 and 4 years of age. ○—•—○ 55 dB; ●–● 70 dB; △--△ 85 dB.

slope and peak magnitudes. Furthermore, the primary slope, secondary slope, and total response times were all significantly shorter in the older ages.

In this same study we also explored the extent to which relative reactivity was stable across the different age periods. The results failed to demonstrate predictability from the newborn period to the 2½- and 5-month ages. However, in general there was a considerable degree of predictability from the 2½- to the 5-month period.

## MONOPHASIC OR DIPHASIC RESPONSE

At first it would appear reasonable to offer the generalization that the newborn response to a nonsignal sudden stimulus is monophasic acceleration whereas by 2½ and 5 months of age the response is triphasic. This would suggest that, in some manner, age-related factors alter the basic *form* of the cardiac response in two qualitative aspects: the initial and secondary decelerative phases. However, it should be remembered that thus far we have been dealing with the *average* cardiac response curves. Closer examination of the individual trial response curves reveals that, under certain conditions, the newborn *is* capable of generating a secondary decelerative response.

In a previously published paper (Richmond, Lipton, & Steinschneider, 1962) we suggested an analytic approach for the purpose of studying the neonate's

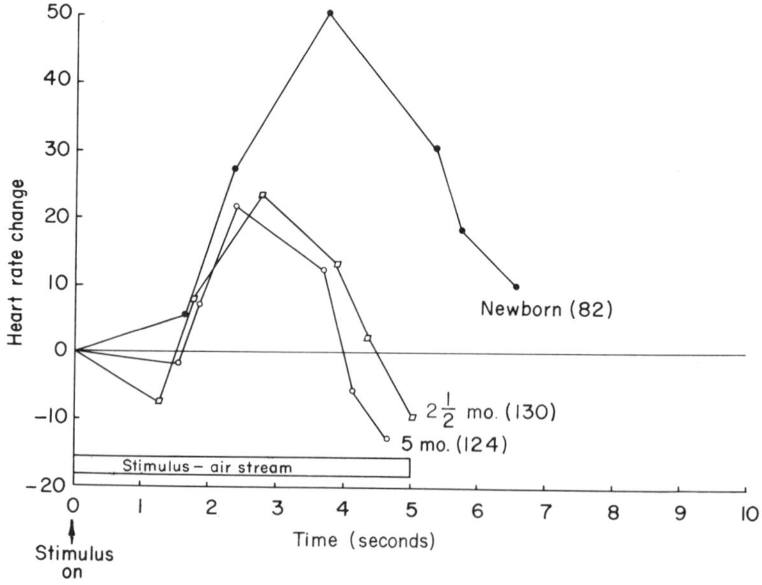


**Fig. 7.** Average cardiac rate response curves derived from infant Mi at three different age periods. (From Lipton *et al.*, 1966.)

capacity to recover from the initial *accelerative* phase of the response. This procedure was limited to those trials wherein the initial cardiac response was an increase in heart rate and considered the relationship between the departure score and the peak magnitude. Figure 8 presents the scatterplot of these two scores for a single subject. Analysis of these scores indicated a statistically significant positive correlation between the departure and peak magnitude scores. The regression of departure score on peak magnitude is depicted by the solid line. Of significance for this discussion is the observation that small peak magnitudes are associated, in general, with negative departure scores. Furthermore, positive departure scores were noted to occur following large peak magnitudes. In this original report we presented data on sixteen newborns. For each subject the slope of this regression line was positive and the $y$-intercept was negative. On the basis of these observations we concluded that, in general, stimulation producing a small initial increase in heart rate also resulted in a cardiac rate return to levels *below* that of the prestimulus. Thus newborns are capable, under certain definable conditions, of producing a biphasic cardiac response to stimulation. It would then seem reasonable to conclude that the return phase of the newborns cardiac response differs quantitatively and not qualitatively from that of the 2½ and 5 month old. Additional support for this conclusion was obtained from data obtained in our brief longitudinal study. Age was not found to influence the slope of the regression of departure score on peak magnitude. However, age did effect significantly the magnitude of the $y$-intercept. The group average $y$-intercept became progressively more negative with increasing age.

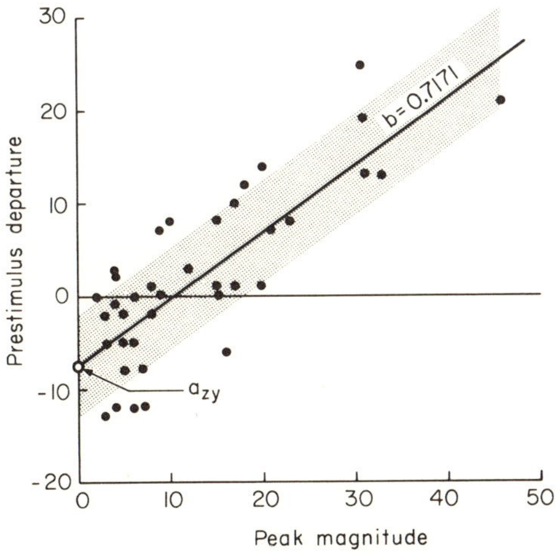


**Fig. 8.** Scatterplot of the prestimulus departure and positive peak magnitude scores obtained from newborn Ch. (Modified from Richmond *et al.*, 1962.)

## INITIAL VALUE

Thus far I have attempted to demonstrate that the cardiac response is influenced by a number of factors: stimulus variables, age-related factors, and individual factors. At this point, I would like to discuss a variable of major significance: one which influences both the magnitude characteristics as well as the primary direction of the response curve. I am referring to the level of organ function at the moment of stimulus presentation.

Wilder (1967) has devoted a considerable amount of effort formulating and reviewing the literature on what he has referred to as the "Law of Initial Value." He has proposed that:

> Given a standard stimulus and a standard period of time, the extent and direction of response of a physiological function at rest depends to a large measure on its initial (pre-experimental) level. The relations are as follows: the higher the initial value, the smaller the response to function-raising, the larger the response to function-depressing stimuli. Beyond a certain medium range of initial values there is a tendency to paradoxic (reversed) responses, increasing with the extremeness of initial values [Wilder, 1967, p. VIII].

It should be emphasized that the "Law of Initial Value" is offered by Wilder as a generalized statistical relationship and makes no claim to suggesting the causal determinants.

It is the purpose of the following discussion to demonstrate that the initial value or prestimulus level is a major determinant of the cardiac response to stimulation, and furthermore, that much can be gained from an understanding of those variables influencing the extent to which the initial value regulates the response to stimulation.

A demonstration of the relevance of the prestimulus heart rate is presented in Figure 9. The data plotted in this figure were obtained from a single newborn exposed repeatedly to a sound stimulus. Peak magnitude is presented on the ordinate and the prestimulus heart rate on the abscissa. Examination of this figure reveals that the peak magnitude is negatively correlated to the prestimulus level. The solid line is the best fit linear regression between these two values. It can be seen from this regression line that, on the average, increasing prestimulus heart rate results in a decrease in the positive value of the peak magnitude. In addition, beyond a certain prestimulus level the peak magnitude, on the average, becomes increasingly negative. In the earlier stages of our research effort it appeared that a better linear regression would be obtained following a log transformation of the prestimulus level. More recently we have reevaluated the importance of this transformation and have found that it adds little over the use of the raw prestimulus values.

A similar within-subject examination was made of each of the magnitude and time response curve measures. On the basis of such an evaluation it was noted that the magnitude response measures were all influenced by the prestimulus

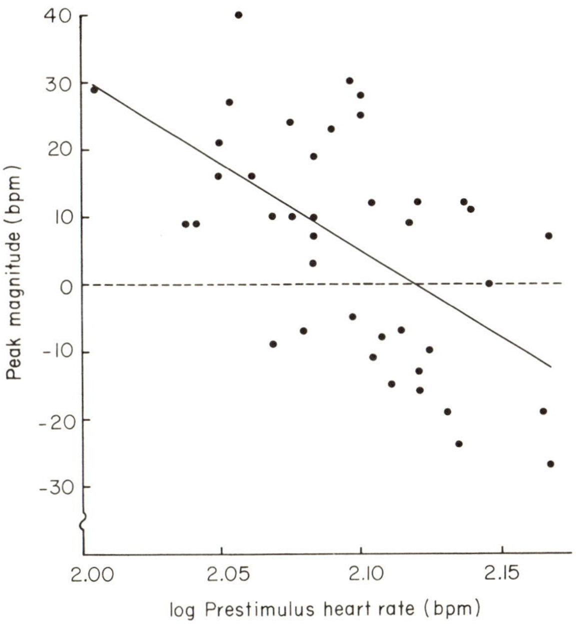


**Fig. 9.** Scatterplot of the peak magnitude and prestimulus heart rate (log) scores obtained from newborn Ca. 55 dB; $y = 548.56 – 258.91(x)$.

level in a manner compatible with the Law of Initial Value. The temporal response measures, however, were not found to be a function of the prestimulus heart rate.

Generally, investigators have viewed the initial value relationship as a hindrance and one to be avoided or statistically adjusted for. However, it is our own belief that in order to understand adequately the cardiac response to stimulation it is absolutely essential that we direct some of our attention to those variables influencing the *extent* to which the initial value exerts a determining role. A similar conclusion was reached by Block and Bridger (1962). In addition it appeared reasonable to hypothesize that individual subjects would differ from one another in the extent to which the initial value influenced the response.

In a paper dealing with methodological considerations (Steinschneider & Lipton, 1965) we suggested that the importance of the initial value could be evaluated by examination of two different measures. The first was the slope of the regression between response magnitude and the initial value. The slope was suggested as a measure of the magnitude affect of changes in the initial value. The correlation coefficient was proposed as still a second measure of the significance of the prestimulus level. It was pointed out that this latter measure would allow for a determination of the extent to which the response was affected by the initial level.

Table 2 contains the group average slope of the regression between *peak magnitude* and prestimulus heart rate in each of four different studies. The values presented in this table as well as those to follow represent the average of the individually determined slopes. A single or double asterisk following the study designation would indicate that the experimental variable was statistically significant. The figures contained within the parentheses were obtained using the raw prestimulus heart rates. All other slopes were calculated from the log transformed prestimulus levels.

Examination of this table reveals that none of the variables under study—intensity, duration, or age—had a statistically significant effect on the slope. There was the suggestion that greater slopes occurred with increasing age from the newborn to 5½ months of age. However, these differences failed to reach statistical significance.

Sound intensity in the newborn period did have a significant effect on the average of the individually determined correlation coefficients (Table 3). In general, larger coefficients were obtained to the more intense stimuli. A comparable study performed on older infants failed to reveal a significant intensity effect. In addition, neither stimulus duration in the neonate nor age appears to influence the correlation between peak magnitude and the prestimulus heart rate.

The relationship between *return magnitude* and prestimulus heart rate was similarly examined. The average slopes are in Table 4 (p. 90). Of considerable importance, age was found to be a statistically significant variable. A unit change in the prestimulus level had a progressively greater effect on the return magni-

TABLE 2
*Peak Magnitude vs. Prestimulus Heart Rate*
*Average Slopes*

| | Experimental condition | | | |
|---|---|---|---|---|
| Study | *A* | *B* | *C* | *D* |
| Intensity-Newborn | −189.8 | −218.6 | −202.6 | −198.6 |
| Duration-Newborn | −148.8<br>(−0.500) | −170.4<br>(−0.615) | −172.2<br>(−0.620) | |
| Longitudinal | −188.8 | −245.5 | −260.0 | |
| Intensity-Infant | −142.9<br>(−0.535) | −129.1<br>(−0.480) | −127.6<br>(−0.477) | |

TABLE 3
*Peak Magnitude vs. Prestimulus Heart Rate*
*Average Correlation Coefficient*

| | Experimental condition | | | |
|---|---|---|---|---|
| Study | *A* | *B* | *C* | *D* |
| Intensity-Newborn* | –.508 | –.604 | –.585 | –.647 |
| Duration-Newborn | –.523 | –.558 | –.585 | |
| Longitudinal | –.726 | –.623 | –.770 | |
| Intensity-Infant | –.397 | –.371 | –.462 | |

*$p < .05$

tude in the 2½- and 5-month old infants. The initial value also was of greater significance in the 5 month old (when compared to the newborn) when the correlation coefficients were evaluated (see Table 5, p. 90).

The stability of the individually determined relationship measures (*b, r*) was tested by calculating the Kendalls Coefficient of Concordance across the experimental variable within each study. These coefficients are in Table 6 (p. 91). Evidence of stability was obtained in all but the longitudinal study. Newborns with relatively larger peak magnitude slopes or correlation coefficients at one intensity level or stimulus duration tended to maintain their same relative position at other intensity levels or durations. Stability of the peak magnitude slopes was also obtained in the Intensity-Infant study. The longitudinal study was the only one failing to result in a significant coefficient of concordance for the peak magnitude slopes. However, a significant Pearson Product Moment Correlation coefficient was obtained between 2½ and 5 months of age. None of the concordance values was significant for return magnitude.

## ACCELERATION OR DECELERATION (CROSSOVER POINT)

In 1959, Bridger and Reiser published their study in which a group of neonates was studied on two successive days and on whom cardiac rate was recorded. They noted that all neonates responded in a manner consistent with the Law of Initial Value. In this study, they explored the crossover point as one of their subject response measures.

> It was observed that most babies have a rather specific initial heart rate value at which they give no heart rate change with stimulation, even though a behavioral response is noted. This value was observed experimentally in most of the infants and was calculated for all the babies by regression analysis and referred to as the crossover point—the prestimulus value at which the regression line crosses the *O* axis [Bridger & Reiser, 1959, p. 270].

TABLE 4
*Return Magnitude vs. Prestimulus Heart Rate*
*Average Slopes*

| | Experimental condition | | | |
|---|---|---|---|---|
| Study | *A* | *B* | *C* | *D* |
| Intensity-Newborn | 126.5 | 150.4 | 135.7 | 77.4 |
| Duration-Newborn | 74.4<br>(0.254) | 107.8<br>(0.388) | 96.6<br>(0.350) | |
| Longitudinal* | 105.7 | 147.6 | 222.7 | |
| Intensity-Infant | 74.7<br>(0.354) | 87.3<br>(0.327) | 65.3<br>(0.228) | |

*$p < .05$

TABLE 5
*Return Magnitude vs. Prestimulus Heart Rate*
*Average Correlation Coefficient*

| | Experimental condition | | | |
|---|---|---|---|---|
| Study | *A* | *B* | *C* | *D* |
| Intensity-Newborn | .273 | .363 | .300 | .188 |
| Duration-Newborn | .282 | .328 | .300 | |
| Longitudinal** | .371 | .273 | .544 | |
| Intensity-Infant | .207 | .217 | .226 | |

**$p < .01$

The individually determined crossover point was evaluated for reliability by correlating the values obtained 24 hours apart. The resulting coefficient was 0.84.

Our own interest in the crossover point derives not only because of its potential for reflecting individual differences. When obtained from the peak magnitude vs. prestimulus regression it also defines that prestimulus level above and below which the form of the response curve differs. Stimulation at a time when the prestimulus heart rate level is smaller than the crossover point will result, on the average, in an initial increase in heart rate, whereas stimuli occurring at a time when the prestimulus heart rate is greater than the crossover point will produce a decelerative response. It appeared that we might better understand the initial directional aspect of the response curve to stimulation by inquiring into the effect of a number of variables on the crossover point as determined from the peak magnitude vs. prestimulus level regression.

Table 7 contains the cardiac rate group average crossover points for the experimental variables in each of four studies. The values in this table were obtained by calculating the average of the individually determined crossover points. Neither stimulus duration (newborn) nor age (longitudinal) appeared to affect these values. However, stimulus intensity was a significant variable for both newborns and the infants between 1 and 4 years of age. In both studies intensity had the same effect Increasing the intensity of stimulation resulted in a larger crossover point.

The significance of this observation can be seen in Figure 10 which presents the regression lines derived for newborn *Ca* at each of the four sound intensities. The crossover point for each regression line would be that prestimulus heart rate for which peak magnitude is zero. Consistent with the averages for the group, increasing stimulus intensity is associated with a shift of the crossover point to the right. At a log prestimulus heart rate of 2.0 (100 bpm) each stimulus intensity produced an average positive peak magnitude response, indicating that cardiac rate acceleration was the initial response to stimulation for all intensities. If we now examine the average peak magnitude at 2.15 (or 133 bpm) we see a somewhat different relationship. The two loudest stimuli (85 and 100 dB) continued to produce, on the average, a positive peak magnitude. However, the

TABLE 6

*Coefficient of Concordance (W) Heart Rate*

| | *b* | | *r* | |
|---|---|---|---|---|
| Study | Peak mag | Return mag | Peak mag | Return mag |
| Intensity-Newborn | 0.690** | 0.440 | 0.473* | 0.317 |
| Duration-Newborn | 0.910** | 0.417 | 0.735* | 0.418 |
| Longitudinal | 0.485 | 0.417 | 0.322 | 0.389 |
| Intensity-Infant | 0.640* | 0.375 | 0.444 | 0.260 |

$*p < .05$
$**p < .01$

TABLE 7
*Average Crossover Point Heart Rate (bpm)*

| | Experimental condition | | | | Coef. conc. |
|---|---|---|---|---|---|
| Study | *A* | *B* | *C* | *D* | *W* |
| Intensity-Newborn** | 125.7 | 136.9 | 148.3 | 161.4 | 0.685** |
| Duration-Newborn | 141.0 | 138.6 | 141.9 | | 0.794* |
| Longitudinal | 152.7 | 162.5 | 158.1 | | 0.548 |
| Intensity-Infant** | 127.8 | 135.5 | 142.7 | | 0.921** |

$^*p < .05$
$^{**}p < .01$

two weaker stimuli (55 and 70 dB) resulted in a negative peak magnitude. Thus at a prestimulus level of 133 bpm an *accelerative* response resulted from stimulation with the 85 and 100 dB sound whereas the cardiac rate response *decelerated* to the presentation of the 55 and 70 dB sound. When we stimulated at still a higher log prestimulus heart rate (i.e., 2.2, 158.5 bpm) then only the most intense stimulus resulted in acceleration and the other three stimuli produced a decrease in heart rate.

In summary, the monophasic direction of the cardiac rate response to stimulation is dependent upon at least two variables, the prestimulus heart rate and the intensity of stimulation. A further point must be made. Although in the majority of published studies the *average* cardiac rate response of newborns is described as being primarily one of acceleration, the newborn can and will respond to stimulation with deceleration if the testing conditions are appropriate.

The procedure for studying the reliability of individual differences has already been described. The results of this analysis are also presented in Table 7. Except for the longitudinal study all coefficients of concordance were statistically significant.

## BEHAVIORAL STATE

The role of behavioral state in modifying responsiveness to standard stimuli has intrigued a number of investigators. This interest stems from both practical as well as theoretical consideration. For example, Lenard, von Bernuth, and Prechtl (1968) investigated the behavioral state variable to define the optimal

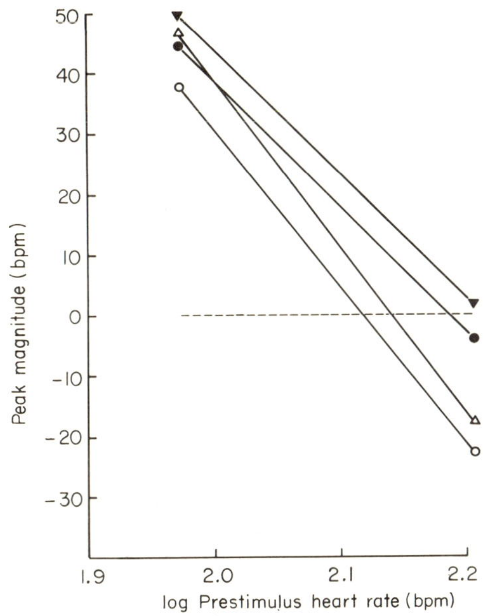


**Fig. 10.** The linear regression lines derived from baby Ca tested with four intensities of white noise. ○ 55 dB; △ 70 dB; ● 85 dB; ▼ 100 dB.

circumstances for the elicitation of a variety of reflexes and "as a tool in the further elucidation of the neurophysiological organization of reflex mechanisms in the human neonate [p. 177]."

Behavioral state has also been suggested as a major variable influencing the directional and some of the quantitative aspects of the cardiac rate response in subjects of all ages. It has been proposed that stimulation presented in the waking state will result in cardiac deceleration whereas stimulation during sleep will produce acceleration. The theoretical significance of this proposal stems from the hypotheses that (*a*) cardiac rate deceleration is a concomitant of efforts to increase "environmental intake" and acceleration to attempts at "rejection of the environment," and (*b*) sleeping is associated with efforts to *exclude* external stimulation and the awake state with efforts to *include* environmental stimulation.

Research efforts attempting to determine the influence of behavioral state on the cardiac rate response have been complicated by the correlation between behavioral state and the unstimulated cardiac rate. Birns, Barten, Cronin, Newton, and Bridger (1967) demonstrated in neonates a progressively increasing mean prestimulus heart rate in association with the state variable ranging from quiet sleep to awake to crying. Lewis, Bartels, and Goldberg (1967) and Lewis, Dodd, and Harwitz (1969), in studies on 2- to 8-week old infants, observed lower unstimulated heart rates in the sleeping state as opposed to the waking state.

Our own data add additional support to this correlation. In several of the studies from our laboratory, behavioral state was determined in the period immediately prior to each stimulus presentation. Since stimulation was not presented during periods of gross movements, vocalization, or crying, we divided state into three categories employing essentially the criteria proposed by Prechtl and Beintema (1964): State 1 (sleep)–eyes closed, no or very slight movements, regular respiration; State 2–eyes closed, no or very slight movements, irregular respirations; State 3 (awake)–eyes open, generally inactive. In our studies no effort was made to ensure that each behavioral state occurred equally often during a given test session.

The group average prestimulus heart rate for the sleep and awake states are presented in Table 8. These averages were obtained by determining initially the within-subject mean heart rate and then calculating the group average. In each of the studies the awake state was associated significantly with a greater prestimulus heart rate. It is important to point out that although the means are different, there was still a considerable amount of overlap in the heart rate distributions for the two states.

As a consequence of the above relationship, evidence supporting the importance of the *state* variable must demonstrate an effect independent of the prestimulus level.

TABLE 8

*Behavioral State and Prestimulus Heart Rate (Average)*

| Study | State | |
|---|---|---|
| | Sleep | Awake |
| Intensity-Newborn* | 113.4 | 125.1 |
| Intensity-Temperature-Newborn | | |
| 75 ° F* | 123.9 | 132.5 |
| 90 ° F* | 126.8 | 133.0 |
| Duration-Newborn* | 114.4 | 130.8 |
| Infant* | 114.7 | 131.7 |
| Intensity-Infant* | 123.5 | 139.3 |

*$p < .05$

### *"Initial Response"*

Lewis *et al.* (1967, 1969) on the basis of two studies performed on infants 2 to 8 weeks of age, concluded that "when asleep, infants show initial cardiac acceleration but when awake show either little response or cardiac deceleration [1969, p. 267]." This conclusion derives primarily from an examination of their "initial response" measure; defined as the difference between the average of the three beats prior to stimulation and the average of the five beats immediately following stimulus presentation. The *awake* "initial response" was not found to differ significantly from the *asleep* "initial response" in their 1967 study. However, a significant difference was obtained in the 1969 study. All of twelve subjects observed in the latter study demonstrated cardiac acceleration during the sleep state, whereas while awake, eight of these subjects "responded" with cardiac deceleration.

A critical examination of these two papers suggested that the conclusions derived from these data had to be viewed with caution for at least three reasons.

1. There was no consistency between the two studies.

2. These same authors were unable to demonstrate a significant difference in the "initial response" between their stimulation and control trials.

3. They failed to consider the possible influence of the prestimulus level on the "response."

Because of the importance of the conclusions reached by Lewis and his colleagues, it seemed appropriate to examine some of our own data for the purpose of evaluating the validity of these concerns. This was accomplished by obtaining the "initial response" for the unstimulated control trials in the Intensity-Newborn study. We then determined the within-subject correlation coefficient and regression coefficient between the "initial response" and the prestimulus level (prestimulation 3-beat average).

In all nine neonates there was a negative relationship between the "initial response" and its prestimulus level (see Table 9). The group average correlation coefficient and regression coefficient were small but significantly different from zero. Furthermore the crossover points derived from these regressions were all, except for one instance, within the range of reasonably normal heart rates. Thus, at low prestimulus heart rates and *in the absence of experimenter-controlled stimulation,* the "initial response" was one of acceleration. At higher prestimulus levels, the "response" was deceleration. These data are interpreted as empirical support for questioning the validity of the conclusions derived from the "initial response" measure.

### *Z Transformation*

Previously, we discussed the influence of a number of variables on the within-subject regression of peak and return magnitudes on prestimulus heart

TABLE 9
*"Initial Response" Control Trials*

| Subject | $b_{yx}$ | $r$ | C.O.P. |
|---|---|---|---|
| Ca | −0.288 | −0.509 | 127.9 |
| Co | −0.100 | −0.277 | 132.0 |
| Gr | −0.201 | −0.395 | 130.4 |
| Ha | −0.088 | −0.225 | 128.2 |
| Ma | −0.112 | −0.228 | 130.1 |
| Mc | −0.159 | −0.390 | 117.3 |
| Re | −0.150 | −0.237 | 103.4 |
| Sa | −0.012 | −0.039 | 195.1 |
| Un | −0.345 | −0.766 | 116.4 |
| $\bar{X}$ | −0.162** | −0.371** | 131.2 |

$^{**}p < .01$

rate. In the determination of these regression lines, all stimulus trials were included regardless of the behavioral state prior to the onset of stimulation. A review of each within-subject regression revealed that each of the three states was represented in varying frequency for most of the subject.

The following procedure was employed to assess the state variable. All subjects were included in whom there were at least sleep and awake stimulus trials. Each obtained magnitude score was transformed to a $Z$-score by subtracting it from the score predicted from the appropriate within-subject regression line and dividing this difference by the appropriate within-subject standard error of estimate (the variability around the regression line). The resulting $Z$-scores were then divided into three subcategories on the basis of behavioral states. Only the sleep ($S$) and awake ($A$) states were subjected to further analysis. This analysis consisted of treating all the $Z$-scores within a given state as if they were derived from a single population of $Z$-scores and a group average obtained. These averages are presented in Table 10. No effect of state would result in average $Z$-scores approximating zero. If awake was associated with an initial decrease in heart rate and the sleep state with acceleration, then the average $Z$-score would be negative and positive, respectively. The observed mean difference between states was subjected to a $t$-test of statistical significance.

The analysis of the time measures was simplified because of their lack of correlation with the prestimulus heart rate. However, the same basic statistical approach was employed and each obtained time score was transformed to a $Z$-score by means of the formula: raw score-average/standard deviation. The group means are also contained in Table 10.

In none of the newborn studies were the magnitude measures significantly affected by behavioral state. This contrasts with our results employing subjects between the ages of 1 and 4 years. In both such studies the awake peak magnitudes were on the average significantly *greater* than the sleep peak magnitudes. These results would indicate that once correcting for the initial value, the awake state resulted in a *greater* initial accelerative response or a smaller initial decelerative response when compared to the sleep state.

No consistent significant differences were obtained for the time measures. Statistically significant longer peak times were obtained for the awake state in one of the newborn studies and significantly longer return times for the sleep state in a different newborn study. Because the remaining newborn studies failed to yield similar results, the two significant differences should be viewed with caution.

In summary, the results of this analysis would indicate that behavioral state is not of significance in the newborn period. However, state does appear to influence the cardiac rate response at a later age. Furthermore, the direction of this effect is for *greater* positive peak magnitudes to occur while awake.

This conclusion is contrary to that reached by Lewis and his colleagues (Lewis *et al.,* 1967, 1969). The discrepancy could be due to differences in the response measures that were considered. In addition, Lewis *et al.* failed to correct for the effect of prestimulus level when considering the magnitude measures. The present conclusions also differ, in part, from that of Graham and Jackson (1970). Graham and Jackson, discussing the results obtained from a number of unpublished studies, reach the conclusion, similar to our own, that the effect of state appears to be related to developmental processes. However, we differ in the directional aspect of the effect of state. Similar to Lewis *et al.,*

TABLE 10

*Effect of Behavioral State (Heart Rate Average Z)*

| Study | Magnitude | | | | Time | | | |
|---|---|---|---|---|---|---|---|---|
| | Peak | | Return | | Peak | | Return | |
| | *A* | *S* | *A* | *S* | *A* | *S* | *A* | *S* |
| Intensity-Newborn | .012 | –.116 | .259 | –.005 | .087 | –.140 | –.078 | –.037 |
| Intensity-Temperature-Newborn | | | | | | | | |
| 75 dB | .097 | –.041 | –.018 | .041 | .091 | .022 | –.119 | .001 |
| 90 dB | .032 | –.042 | –.033 | –.083 | .196* | –.243 | .124 | .069 |
| Duration-Newborn | –.162 | –.099 | .187 | –.015 | .167 | –.093 | –.287* | .162 |
| Infant | .241* | –.136 | –.080 | –.035 | .004 | –.024 | .066 | .033 |
| Intensity-Infant | .229* | –.123 | –.011 | .009 | .179 | –.064 | –.130 | .026 |

$^*p < .05$

Graham and Jackson suggest that a decelerative response is more apt to occur in the waking state. Once again we noted a greater tendency toward acceleration. A number of differences between the studies they report and our own could account for the different conclusions, some of which are methodological and others analytic. To single out but one of the latter, in the studies reported by Graham and Jackson a common linear regression was used in all subjects to adjust or correct for the prestimulus level. Support for this procedure would derive from the study by Clifton and Graham (1968) in which newborns were tested over a 5-day period. Examination of the within-subject regression failed to reveal statistical evidence of individual differences. However, the data presented by Bridger and Reiser (1959), as well as those discussed earlier, would indicate that regression coefficients are individually determined. If we accept the latter conclusion, then the use of a common regression line for all subjects would not adequately adjust or correct for the initial level. As a consequence, the results discussed by Graham and Jackson might still be a reflection of the prestimulus level and not the effect of behavioral state.

## GENERAL SUMMARY

In the course of this presentation we have attempted to examine the cardiac rate response to a nonsignal stimulus by focusing primarily on the individual infant. As a means of accomplishing this, considerable effort was expended in the development of an analytic procedure that would maintain the integrity of each response trial. Employing this approach, it was possible to investigate in some detail some of the factors influencing the average response as well as the sources of some of the variation between trials.

On the basis of this examination, it is reasonable to conclude that the cardiac response curve in a neonate can assume a number of different forms: monophasic acceleration, monophasic deceleration, and biphasic (acceleration–deceleration). It is also possible to specify some of the conditions predisposing to the development of one or the other response type. At low prestimulus levels, monophasic acceleration is the most likely response. Increasing the prestimulus level would result in a shift in the form of the response toward the biphasic type. Further increases would be associated with a monophasic-deceleration type of curve. In essence, we are proposing that there is a continuum of response shapes influenced by the prestimulus level and anchored at one end by a monophasic-acceleration curve and a monophasic-deceleration curve at the other with the biphasic curve in between. Our results on the effect of stimulus intensity would suggest that weak stimuli act to shift the curve shape toward the monophasic deceleration end and intense stimuli toward the monophasic acceleration end of the continuum. The influence of age can also be viewed within this same conceptual scheme. Increasing age influences not only the quantitative features of the response but also shifts the response curve in the direction of deceleration. It would also be interesting to speculate that an increase in the level of arousal (going from sleep to an awake state), in the non-neonate, is comparable to increasing the intensity of stimulation. Individual differences exert a significant influence on the magnitude and temporal characteristics of the response as well as the point on the prestimulus continuum at which there is an alteration in the shape of the response curve.

It is important to point out that the demonstration of individual differences in the neonate does not necessarily imply that we are dealing with genetic differences. The individuality of responsiveness noted in the neonate could also be the result of intrauterine experiences, or differential variables operating at the time of delivery. Admittedly, it is somewhat disappointing that responsiveness in the newborn period is not predictive of responsiveness at a later age (i.e., 2½ or 5 months). However, this is not surprising when we recognize that cardiac tissue as well as cardiac control mechanisms are undergoing a number of changes during the first few months of life (James, 1968). Furthermore, there is no reason to expect that the rate at which these changes occur will be the same for all individuals. It is more likely that individuals differ along this dimension as well. It is also important not to lose sight of the observation that responsiveness at 2½ months was predictive of the 5-month response.

There is one aspect of the cardiac response at the older ages that appears to be qualitatively different from that of the newborn. Infants between the ages of 2½ months and 4 years were observed to develop a brief deceleration phase prior to the acceleration and subsequent deceleration phases. A similar phase was not obtained in the newborn period. Several studies performed on adult subjects have also presented data demonstrating this initial brief deceleration (Raskin, Kotses, & Bever, 1969a, 1969b; Smith & Strawbridge, 1969). In addition, it would appear from these studies that this decelerative phase is most apt to occur to weak stimuli.

Several studies performed on adults in which respiratory activity has been recorded have demonstrated that the onset of a weak stimulus generally results in a brief respiratory pause (Poole, Goetzinger, & Rousey, 1966; Rousey, Snyder, & Rousey, 1964). These observations suggest the possibility that the initial brief cardiac rate deceleration might be secondary to the respiratory response. In apparent contradiction to this hypothesis is the observation that neonates also show a similar respiratory pause without, at the same time, demonstrating the bradycardia (Steinschneider, 1968). However, this contradiction can be understood on the basis of those data which indicate that the interrelationship between respiratory and cardiac activity increases initially with age (Reeve & DeBoer, 1960).

There is developing a considerable amount of literature dealing with the probable significance of the cardiac rate response to stimulation. Much of this discussion is formulated within the conceptual framework suggested by Sokolov (1963). Sokolov has proposed a number of functional response systems, two of which—the orientation reflex and the defense reflex—are of particular interest. The orientation reflex is characterized as a nonspecific set of responses elicited by stimulus change and which rapidly habituates with stimulus repetition. In contrast, the defense reflex is elicited by intense stimulation and is not diminished by repeated presentation of the stimulus. A proposed consequence of the orientation reflex is an increased sensitivity to stimulation whereas the effect of the defense reflex would be "... the breaking away from or limitation of the activity of the stimulus [Sokolov, 1963, p. 14]." Many of the studies exploring the significance of the cardiac rate response to discrete stimuli have attempted to determine which component or components of the response curve best fit the orientation or defense patterns (Clifton & Graham, 1968; Clifton & Meyers, 1969; Germana & Klein, 1968; Lang & Hnatiow, 1962; Meyers & Gullickson, 1967; Raskin, Kotses, & Bever, 1969a, 1969b; Smith & Strawbridge, 1969). These studies have approached this problem by varying stimulus properties and by investigating the effect of repeated stimulation. In general, the results obtained from such studies are compatible with the hypothesis that bradycardia is a component of the orientation reflex. However, these data can only be viewed as indirect evidence of the *functional* implications of cardiac rate changes. What is clearly needed are studies which are more direct in their approach.

One such methodological approach would place emphasis on the proposed relationship between the two response systems and sensitivity to stimulation. If the deceleration noted in the cardiac response curve is part of the orientation reflex, then we can anticipate an increase in stimulus sensitivity at the time of its occurrence. Conversely, if acceleration is a concomitant of the defense reaction, then stimulus sensitivity should decrease at these times. Furthermore, there should be a correlation between the degree of cardiac rate change and change in sensitivity. In general, it should be possible to fractionate experimentally the complex cardiac response curve by determining the direction of sensitivity changes at various points on the response curve. Studies employing this approach have yet to be done.

## REFERENCES

Birns, B., Barten, S., Cronin, V., Newton, L., & Bridger, W. H. State as an experimental variable in psychophysiological studies of the neonate. Paper presented at the annual meeting of the American Psychosomatic Society, New Orleans, April 1967.

Block, J. D., & Bridger, W. H. The law of initial value in psychophysiology: a reformulation in terms of experimental and theoretical considerations. *Annals of the New York Academy of Sciences,* 1962, vol. 98, 1229-1241.

Bridger, W. H., & Reiser, M. F. Psychophysiologic studies of the neonate: an approach toward the methodological and theoretical problems involved. *Psychosomatic Medicine,* 1959, vol. 21, 265-276.

Clifton, R. K., & Graham, F. K. Stability of individual differences in heart rate activity in the newborn period. *Psychophysiology,* 1968, vol. 5, 37-50.

Clifton, R. K., & Meyers, W. J. The heart-rate response of four-month-old infants to auditory stimuli. *Journal of Experimental Child Psychology,* 1969, vol. 7, 122-135.

Crowell, D. H., Davis, C. M., Chun, B. J., & Spellacy, F. J. Galvanic skin reflex in newborn humans. *Science,* 1965, vol. 148, 1108-1111.

Germana, J., & Klein, S. B. The cardiac component of the orienting response. *Psychophysiology,* 1968, vol. 4, 324-328.

Graham, F. K., & Clifton, R. K. Heart-rate change as a component of the orienting response. *Psychological Bulletin,* 1966, vol. 65, 305-320.

Graham, F. K., & Jackson, J. C. Arousal systems and infant heart rate responses. In L. P. Lipsitt & H. W. Reese (Eds.), *Advances in child development and behavior.* Vol. 5. New York: Academic Press, 1970. Pp. 69-117.

James, T. N. Sudden death in babies: new observations in the heart. *American Journal of Cardiology,* 1968, vol. 22, 479-506.

Lacey, J. I. Psychophysiological approaches to the evaluation of psychotherapeutic process and outcome. In E. A. Rubinstein & M. B. Parloff (Eds.), *Research in psychotherapy.* Washington, D. C.: American Psychological Association, 1959. Pp. 160-208.

Lacey, J. I., Kagan, J., Lacey, B. C., & Moss, H. A. The visceral level: situational determinants and behavioral correlates of autonomic response patterns. In P. H. Knapp (Ed.), *Expression of the emotions in man.* New York: International Universities Press, 1963. Pp. 161-196.

Lang, P. J., & Hnatiow, M. Stimulus repetition and the heart rate response. *Journal of Comparative and Physiological Psychology,* 1962, vol. 55, 781-785.

Lenard, H. G., von Bernuth, H., & Prechtl, H. F. R. Reflexes and their relationship to behavioral state in the newborn. *Acta Paediatrica Scandinavica,* 1968, vol. 57, 177-185.

Lewis, M., Bartels, B., & Goldberg, S. State as a determinant of infants' heart rate response to stimulation. *Science,* 1967, vol. 155, 486-488.

Lewis, M., Dodd, C., & Harwitz, M. Cardiac responsivity to tactile stimulation in waking and sleeping infants. *Perceptual and Motor Skills,* 1969, vol. 29, 259-269.

Lipton, E. L., & Steinschneider, A. Studies on the psychophysiology of infancy. *Merrill-Palmer Quarterly,* 1964, vol. 10, 103-117.

Lipton, E. L., Steinschneider, A., & Richmond, J. B. Autonomic function in the neonate: III. Methodological considerations. *Psychosomatic Medicine,* 1961, vol. 23, 461-471.

Lipton, E. L., Steinschneider, A., & Richmond, J. B. Autonomic function in the neonate: VII. Maturational changes in cardiac control. *Child Development,* 1966, vol. 37, 1-16.

McDonald, D. G., Johnson, L. C., & Hord, D. J. Habituation of the orienting response in alert and drowsy subjects. *Psychophysiology,* 1964, vol. 1, 163-173.

Meyers, W. J., & Gullickson, G. R. The evoked heart rate response: the influence of auditory stimulus repetition, pattern reversal, and autonomic arousal level. *Psychophysiology,* 1967, vol. 4, 56-66.

Poole, R., Goetzinger, C. P., & Rousey, C. L. A study of the effects of auditory stimuli on respiration. *Acta Oto-Laryngologica,* 1966, vol. 61, 143-152.

Prechtl, H. F. R., & Beintema, D. *Clinics in developmental medicine.* No. 12. *The neurological examination of the full term newborn infant.* London: Heinemann, 1964.

Raskin, D. C., Kotses, H., & Bever, J. Autonomic indicators of orienting and defensive reflexes. *Journal of Experimental Psychology,* 1969, vol. 80, 423-433. (a)

Raskin, D. C., Kotses, H., & Bever, J. Cephalic vasomotor and heart rate measures of orienting and defense reflexes. *Psychophysiology,* 1969, vol. 6, 149-159. (b)

Reeve, R., & DeBoer, K. Sinus arrhythmia: I. Data and patterns from groups of individuals followed from 1 month to 23 years of age. *Pediatrics,* 1960, vol. 26, 402-414.

Richmond, J. B., Lipton, E. L., & Steinschneider, A. Autonomic function in the neonate: V. Individual homeostatic capacity in cardiac response. *Psychosomatic Medicine,* 1962, vol. 24, 66-74.

Roessler, R., Greenfield, N. S., & Alexander, A. A. Ego strength and response stereotypy. *Psychophysiology,* 1964, vol. 1, 142-150.

Rousey, C., Snyder, C., & Rousey, C. Changes in respiration as a function of auditory stimuli. *Journal of Auditory Research,* 1964, vol. 4, 107-114.

Smith, D. B. D., & Strawbridge, P. J. The heart rate response to a brief auditory and visual stimulus. *Psychophysiology,* 1969, vol. 6, 317-329.

Sokolov, Y. N. *Perception and the conditioned reflex.* New York: Macmillan, 1963.

Steinschneider, A. Sound intensity and respiratory responses in the neonate: comparison with cardiac rate responsiveness. *Psychosomatic Medicine,* 1968, vol. 30, 534-541.

Steinschneider, A., & Lipton, E. L. Individual differences in autonomic responsivity: problems of measurement. *Psychosomatic Medicine,* 1965, vol. 27, 446-456.

Steinschneider, A., Lipton, E. L., & Richmond, J. B. Auditory sensitivity in the infant: effect of intensity on cardiac and motor responsivity. *Child Development,* 1966, vol. 37, 233-252.

Wilder, J. *Stimulus and response: The law of initial value.* Bristol: Wright, 1967.

## DISCUSSION

The discussion of Dr. Steinschneider's presentation centered primarily around methodological questions and the need for additional research. In dealing with cardiac acceleration and deceleration, many variables must be considered, and it is often difficult to determine which ones are important in relation to the particular response under study. The participants raised many questions which need to be explored further.

Are there important differences in the responses of the two sexes? Are females more stable over time, as some investigators have suggested?

Do behavioral measures such as adaptability tie in with heart rate measures? Do changes in the heart rate of a placid child follow the same pattern as those of an anxious child?

How do immediate behaviors in response to stimuli differ in children of different ages, temperaments, or sexes? What is the significance of these differences in response? For example, does a newborn perhaps react less vigorously to auditory stimuli than an older child because his buffering in the auditory system is greater than that of an older child so that the stimulus has less psychophysiological meaning for him?

Such factors as the threshold for sound must be considered, Dr. Steinschneider replied. Newborns did not show significant cardiac responses to a 55 dB sound. We might interpret this as meaning that this sound level is below threshold for newborns, and therefore does not function as a stimulus for them.

The age of the subject affects the response that is observed in several ways. One can demonstrate histologically that the heart of a newborn is not the same as the heart of a five-month-old child. Changes take place in the tissue of the organ. We cannot assume, therefore, the predictability of cardiac rate changes in older children from what we observe in newborns. Nor can we assume all infants will develop at exactly the same rate. We are dealing with dynamically changing organisms which show many individual differences.

There is an old generalization that there is an essential instability of many physiological processes in the very young infant, and that these tend to stabilize in a matter of a few weeks or months, a participant said. Is this true, or has the supposed instability been due to observational or instrumental errors? Dr. Steinschneider's data, obtained through computer processing, suggest greater stability than earlier studies have indicated.

Data from some studies seem to suggest that the orienting or decelerating response habituates over time but the defense or accelerating reaction does not. Is this supported by Dr. Steinschneider's data?

It is difficult to tease out habituation, Dr. Steinschneider replied. In his own studies, experimental conditions are not appropriate because multiple stimuli are used in the tests. One can find habituations, however, if the conditions are right. One has to use relatively long duration stimuli and short intertrial intervals. Parametric studies of habituation are needed.

Is there any way to determine the primary effect of a given stimulus, a participant asked. Would it be possible, for example, to get an enzyme or hormonal profile of the plasma which would show that changes occur as a result of the stimulus which lead, in turn, to a succession of additional changes, and ultimately to the final effect on heart rate? What happens between the stimulus and the change in heart rate?

It would appear, another participant suggested, that the single variable type of study may be passé. Dr. Steinschneider has found that cardiac deceleration occurs in some newborn infants, but that it occurs under highly specialized circumstances which necessitate use of a very complex model because multiple interacting variables are simultaneous determinants of the deceleration phenomenon. We must consider not only the relationship between stimulus intensity and response, but also interactions among such variables as basal heart rate at the start of the stimulus trial. It is conceivable, for example, that deceleration occurs only in subjects who have a high heart rate at the time a low intensity auditory stimulus is administered. Cardiac acceleration has been shown to occur even without presentation of a stimulus; it may be that this is also true of deceleration. It is also possible that the whole range of acceleration and deceleration could be demonstrated within the same subject by offering the stimulus to him when his heart rate is at different levels.

How can we be sure the heart rate change in the neonate really has any relevance in either a biological or a behavioral context, another participant asked. It does not correlate with later measures of heart rate; and in any case, it is not safe to generalize from studies of adults back to infants. Where does that leave us? Does the heart rate have any meaning in predicting the infant's behavior?

It is important not to exaggerate the implications of changes in heart rate, Dr. Steinschneider said. This is only one of many physiological changes that take place, but for study purposes we have to single out one of the response organs, and even the end point of that organ system. Even when we do this, it is difficult to identify what produces what because many changes commence almost simultaneously. The causal sequence from input to output is not clear, particularly if efforts to modify physiological responses involve the use of drugs such as atropine which in themselves produce changes.

In general, the acceleration and deceleration responses appear to be a continuum. The newborn is capable of producing monophasic accelerative responses, monophasic decelerative responses, and diphasic responses. These can be looked at from the standpoint of the average curve, or they can be looked at in terms of magnitude of response or the temporal relationships.

It is difficult to determine what are the base levels of functioning independent of stimulation. It is an error to assume that an individual under controlled test conditions is not receiving stimulation, either internal or external. The room is not totally soundproof. There is activity going on within the organism. Stimuli are present, even if we cannot define them; and it may well be that changes take place in response to these unidentified stimuli even in supposedly unstimulated trials. No correlation has been identified between the magnitude scores of stimulated and unstimulated trials in infants. With older youngsters, there is some correlation in magnitude of response, but not in the temporal relationship to the onset of the trials. This is extremely variable.

All we can really do is to set up a controlled condition and determine whether there are differences in terms of the particular variable we have chosen to study. If there is no response, then we can assume that there is no relationship between the properties of the stimulus and the response. If a response occurs when a stimulus is given, we must consider whether the response is due to that particular stimulus, or to something else that is going on at the same time internally or externally.

We need to ask many questions about the relationship of heart rate to behavior, Dr. Steinschneider continued. For example, what does habituation of a cardiac response imply in terms of behavior? What does the orienting reflex mean in terms of sensitivity and possible effects on learning? What does the defense reaction mean in terms of its effects?

Those who are working with measures of this kind start off with a tremendous amount of faith and hope that their findings will prove to be significant in relation to learning and behavior, but no one can state categorically at this stage of research whether this will prove to be true. In a sense, all of the studies of cardiac acceleration and deceleration that have been done thus far have been part of a process of "tooling up" for future questions about the behavioral significance of these responses. We now know the kinds of conditions under which we get certain types of responses. The next question is, what does all of this mean?

## ATTENTION IN THE INFANT: AVENUE TO THE STUDY OF COGNITIVE DEVELOPMENT[1]

*Robert B. McCall*
Fels Research Institute

Currently, there is quite a flurry of interest in the study of the distribution of attention in human infants, where "attention" connotes a selective perceptual-cognitive orientation towards some attribute of the stimulus environment. The focus of this research has been on the parameters of the stimuli and of the infant's experience that govern visual and auditory attention as well as the distribution of his manipulative exploration (i.e., "play").

At least a component of the interest in these topics derives from the potential this area holds for the study of cognitive development. In addition to the sheer pervasiveness of attentional behavior throughout the waking hours of young infants, the allocation of attention to a given stimulus may be propaedeutic to learning about that stimulus event, where "learning" implies the acquisition of a memory for the stimulus. The attentional system acts as a filter on the environment, and those stimuli that pass through can be encoded, remembered, and used to guide future behavior. Consequently, the dispositions of the attentional system are likely to be in harmony with the infant's ability to interact cognitively with the environment, and several authors (e.g., Hunt, 1961,

[1]This research was sponsored by USPHS grant HD 04160-01 to RBM and by USPHS grant FR-05537 to the Fels Research Institute.

1965; Jeffrey, 1968; Piaget, 1952) have emphasized what Hunt (1961, 1965) has called the "match" between what the infant is cognitively capable of ingesting and the potential of the stimulus to service that capability. Presumably, a stimulus event that can provide new information but is not so discrepant from what is familiar that the infant cannot assimilate it will recruit the maximum amount of attention and also be most efficient in promoting accommodative changes in the cognitive system itself. Therefore, a common mechanism may underlie what captivates an infant's attention and what facilitates cognitive development.

Two collateral research efforts have emerged. First, one might study the infant's developing ability to process information as reflected in his increased attention to progressively more "complex" stimuli. From this behavior, one might be able to infer the emergence of cognitive structures. Numerous studies are reported on the distribution of the infant's attention to stimuli varying in complexity and the changes in such a distribution as the child matures.

A second focus has emphasized the differential attention that infants allocate to familiar versus novel (or "discrepant") stimuli. Presumably, with repeated exposure to a given stimulus, the infant learns something about that event. That is, he develops some type of memory engram for the stimulus, and as this engram matures, the stimulus progressively loses its potency to recruit attention. If a new stimulus, unfamiliar to the infant, is then introduced, it should receive more attention than the old familiar one. The study of attention to novelty has several kinds of implications for cognitive development. First, in order for the basic cognitive processes of assimilation and accommodation (Piaget, 1952) to take place, the infant must be able to detect changes in the environment and recognize new stimuli as being unfamiliar to him. Second, the progressive decline in attention to the repeated presentation of a given stimulus (i.e., habituation) *may* reflect the process of acquiring a memory engram for that stimulus and thus constitute an index of the progress of certain kinds of perceptual learning. Third, if the response to a new stimulus is partially a function of the infant's ability to encode and retrieve a memory for a familiar stimulus, then perhaps the distribution of attention to familiar and novel stimuli can form the basis of a method for assessing the parameters and efficiency of rudimentary perceptual learning in nonverbal infants, a rather exciting possibility.

This paper represents a progress report on these general endeavors. A sampling of research findings will be presented along with discussions of some of the issues, problems, and implications of these approaches to studying cognitive development.

## THE MEASUREMENT OF ATTENTION

Although researchers generally agree that some index of ocular orientation to the stimulus (e.g., visual fixation time) is an appropriate measure of attention to visual stimuli, there is little communality of opinion on specific variables. "Attention" is certainly a cluster concept, and one dependent variable may reflect one aspect of it and another measure quite a different facet.

### *Measures Employed*

Visual stimuli are most commonly used, and the length of time an infant looks at a stimulus possesses considerable intuitive validity as a measure of attention. However, there are several methods of measuring looking time. The index most sensitive to stimulus differences appears to be the length of the first look ("first fixation") during a given stimulus presentation (Kagan, Henker, Hen-Tov, Levine, & Lewis, 1966; Lewis, Kagan, & Kalafat, 1966c) rather than total fixation, the sum of all looks during a presentation. Cohen (1969b) has suggested that if the presentation interval is sufficiently long, total fixation may be broken down into two components, one sensitive to the power of the stimulus to attract the infant's gaze away from some other stimulus (e.g., the latency to the first look, the number of looks during a stimulus presentation) and one that reflects the extent to which the infant remains looking at the stimulus once he fixates on it (e.g., the length of the first fixation, the average duration of a fixation). He proposes that stimulus attributes such as size, brightness, and movement govern the latency to the first look at a stimulus or the number of looks it receives in a given period of time while the informational characteristics that the stimulus possesses for the infant determine the first fixation and average duration measures. Cohen's data (1969b) support such a view.

Regrettably, the researcher's concern for homogeneous surroundings within the experimental context has left the infant with very little to look at other than the target stimulus. The consequence is that some of the time the infant spends fixating the stimulus is "blank looking," relatively devoid of cognitive interaction with the stimulus. In the search for a more sensitive index of attention and cognitive processing, the Russian work on the orienting reflex (e.g., Sokolov, 1963) and the psychophysiological research of the Laceys' (Lacey, Kagan, Lacey, & Moss, 1963; Lacey, 1967) have pointed to using heart rate deceleration as a measure of attention. The application of cardiac change to this context has been reviewed by Graham and Clifton (1966), and several studies with infants have demonstrated that cardiac deceleration during a visual fixation is more sensitive to stimulus differences than is looking time (Kagan *et al.,* 1966; McCall & Kagan, 1967b; McCall & Melson, 1969, 1970b). Further, there is no postural response comparable to looking when auditory stimuli are used. However, cardiac change also can be employed with auditory stimuli, thus covering both the visual and auditory modalities with a single dependent measure (e.g., Kagan & Lewis, 1965; Kagan, 1968; Melson & McCall, 1970).

In addition to the above, pupil dilation (Fitzgerald, 1968), sucking suppression (Haith, 1966), and motor quieting (Haith, Kessen, & Collins, 1969) have been used as indices of attention. Further, Siqueland (1969) has made the presentation (or withdrawal) of the target stimulus contingent upon the infant's high amplitude sucking and used the number of sucks as a measure of attention. Smiling and vocalizing have also been tabulated as concomitants of attentional behavior (e.g., Kagan *et al.,* 1966; Kagan, 1968; McCall & Kagan, 1970).

When the child is allowed to interact motorically with the stimuli, the nature of the child's commerce with an object may be broken down into a variety of kinds of manipulation (e.g., holds, mouths, and manipulates with or without visual regard) and levels of activities (secondary and tertiary circular responses, appropriate behavior, and parallel and integrated play with more than one object). A procedure for recording these variables has been used by McCall and Garratt (1969). However, data by Schaffer and Parry (1969) demonstrate that looking and manipulating are distinct systems, with stimulus differences appearing at 6 months for looking but only at 12 months for manipulating.

*Response Selection*

For visual stimuli, the data favor using first fixation (or average fixation) coupled with cardiac deceleration and/or one of the other variables studied (e.g., pupil dilation, movement or sucking suppression). In research from my own laboratory to be reported in this paper, first fixation and cardiac deceleration ("peak magnitude") will be used.[2] Regrettably, it is not clear under what conditions one should expect stimulus differences to be manifested in the fixation or cardiac measures. There is some support (e.g., Kagan *et al.,* 1966; McCall & Kagan, 1967a, 1967b, 1970) for the notion that stimulus differences will be evident in the cardiac measure only when those stimuli bear some relationship to well familiarized events (e.g., faces, discrepancies from stimuli experimentally made familiar over a long term), but this has not always been the case (e.g., McCall & Melson, 1970a).

[2]Cardiac deceleration as used in this paper is most comparable to what Dr. Steinschneider calls "peak magnitude." The context in which this deceleration occurs is somewhat different in our research than in his. For example, we rarely see a pronounced acceleration. Further, Dr. Steinschneider reports rather high correlations (.50–.70) between prestimulus level and deceleration; we do not observe this relationship. The *r* is a nonsignificant .20. These differences probably result from using patterned stimuli that occur over a period of time.

Researchers in the area are plagued by the lack of a theory of the response dynamics of attentional behavior. The concept of attention is not unitary, and the several measures of attention are certainly not comparable. If one is interested in understanding the stimulus side of attentional behavior, as is the emphasis of this paper, one has two choices. First, one can decide not to pursue this research until it is clear just what one or several dependent variables really mean. Conversely, one can assume that a construct validity develops for a measure as its network of relationships with other events (including stimulus parameters) matures. Consequently, one continues with his work, harboring a faith rooted in whatever evidence is available that it will someday be possible to integrate these data and that one's own efforts will contribute to developing that nomological network.

## COMPLEXITY, CONTOUR, RATE OF CHANGE, AND AGE

Some of the first work in the area was an attempt by Fantz (1958, 1961, 1963) and Berlyne (1958) to specify the physical attributes of stimuli that would attract an infant's attention. Fantz demonstrated that the newborn infant attended to patterned more than homogeneous stimuli. Berlyne (1958) suggested that young infants (2–4 months of age) may attend to complex patterns more than simple ones. More recent work has shown that the newborn attends differentially to (*a*) separate aspects of simple visual stimuli (Salapatek, 1968; Salapatek & Kessen, 1966), (*b*) different rates of change and variability of the paths traced by moving lights (Cohen, 1969a; Haith, 1966; Haith *et al.*, 1969), and (*c*) intermittant versus continuous tones (Eisenberg, Griffin, Coursin, & Hunter, 1964).

Most interest has been focused on the role of "complexity" in visual pattern, but on the surface of it the data are not very consistent. Berlyne (1958) found increments of visual fixation for more complex members of some pairs of stimuli but not others. Hershenson, Munsinger, and Kessen (1965) presented neonates with solid, black randomly generated polygons of 5, 10, or 20 sides and observed an inverted-U relationship between complexity and choice based upon relative fixation time in a paired comparison task, while Hershenson (1964), also studying newborns, found decreasing visual attention with increasing complexity defined in terms of the grain of black-white checkerboard patterns. The apparent inconsistency in results suggests that other variables are involved.

### *Contour*

One possibility is that complexity is a cluster concept and that some aspect, such as the length of black-white contour, is the salient variable. Spears (1964) had suggested that contour might be important and Salapatek and Kessen (1966) reported that newborns appeared to fixate their gaze on the vertices of triangles, the points of highest black-white contrast (but see Salapatek, 1968).

Pursuing these leads, McCall and Kagan (1967b) noted that the contour length in the random shapes and checkerboard patterns often used in this type of investigation varies directly with the number of sides or checks. McCall and Kagan (1967b) manipulated *mean* contour length and number of sides in a set of random shapes and found a rather unsystematic interaction between these variables with 4-month old infants. These results were more parsimoniously described as an inverted-U function of *total* contour length.

McCall and Melson (1970a) attempted to determine if stimuli that varied in complexity, but not in contour or area, could provoke differential amounts of attention from 5-month infant girls. Each stimulus was composed of nine white squares outlined in black, and complexity was defined by the amount of asymmetry and irregularity of arrangement of the nine squares on a white background. No attentional differences were observed. In a second study, complexity as defined above was held constant but the amount of area and contour length was varied by enlarging or reducing the size of the squares and the black border on each. Under these conditions, differential attention was paid as a function of both contour and area.

Since this study was performed, Karmel (1969) varied the total contour length in visual patterns independently from adult-rated complexity and found attention to be a function of contour rather than complexity. In a series of studies, Moffett (1969) tested 9- to 19-week old infants with a variety of stimuli composed of different numbers of black lines and line intersections. She concluded that the number of lines (and thus contour length) as well as the number of white rectangles produced by line intersections governed the fixation behavior of her subjects. The suggestion that open white spaces determine attention is interesting in view of Salapatek's (1968) observation that neonates tend to fixate on the middle of white (or black) objects placed against a black (or white, respectively) background.

These several studies agree in assigning functional significance to contour in the deployment of visual attention, but they do not appear to agree on the form of that relationship.

### *Age X Contour*

It may be that these results can be subsumed by an age X contour (complexity) interaction. Brennan, Ames, and Moore (1966) used checkerboards of various grains and demonstrated that young infants (3 weeks old) visually fixated the simple more than the complex patterns, whereas this trend began to reverse itself at 8 weeks of age and by 14 weeks the most complex stimulus received the most attention.

Karmel (1969) integrated his data on 13- and 20-week old infants with that of Brennan *et al.* (1966) and Hershenson (1964) around the square root of the total black-white contour in the patterns. His curves are presented in Figure 1. Karmel proposes that an inverted-U relationship exists between contour and attention but that such a curve moves up the scale of contour and becomes more platykurtic with age.

The length of contour does not possess as much intuitive validity as a measure of complexity as does the number of sides or checks. For example, one could take a circle and enlarge or reduce its size thus varying contour length, but we would hardly call this a manipulation of complexity. Karmel (1969) suggests that as the infant scans a pattern, the amount of contour in that form will relate to the number of on-off brightness changes that impinge upon the retina. Thus, for young infants, "complexity" may actually be mediated by a "rate of change" in stimulation to the visual system.

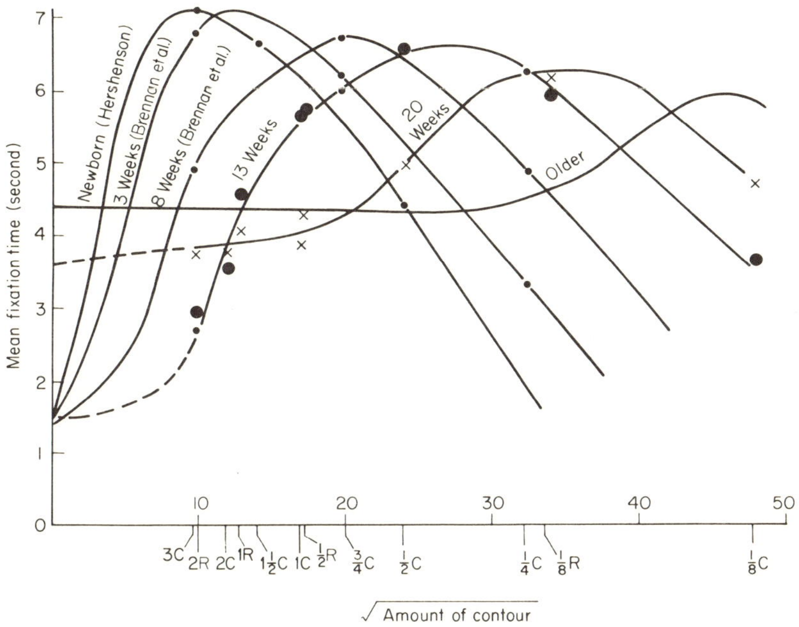


**Fig. 1.** Mean looking time for each stimulus plotted with respect to the square root of the amount of contour contained in the stimulus for different aged *S*s. The letters (*C*) and (*R*) represent redundant and random patterns, respectively. Number prefix indicates size of element used for construction. (Example: ½*C* represents the ½-inch redundant pattern.) A (*c*) (lower case) represents the redundant character of elements of patterns used by Brennan *et al.* (1966) or Hershenson (1964). The number preceding the (*c*) would be equivalent to the check size used by these investigators. (From Karmel, 1969; reproduced by permission of author and publisher.)

*Discordant Notes*

The above analysis, particularly the age X contour interaction, has the power of integrating several diverse findings as well as possessing a certain amount of intuitive appeal. Unfortunately, not all the data are concordant with it. The findings of Brennan *et al.,* (1966), an important contribution to the theoretical curves drawn by Karmel (1969), have not been replicated by Horowitz and Paden (1969). However, Greenberg and Weizmann (Weizmann, personal communication) have data that do support Brennan *et al.* Further, Munsinger and Weir (1967) did not detect age changes in looking to random shapes (presumably also varying in contour), though the ages studies (9–41 months) are older than those used by Karmel (1969). This may merely suggest that age changes in perceptual strategy for this type of stimulation level off some time after 5 months.

Other data indicate that absolute length of contour may not be the only characteristic operating in the traditional manipulations of "complexity." Cohen (1969b) reports the use of checkerboards of two grains (2 vs. 32 black squares) and magnified or reduced the general size of the stimuli. While the manipulation of number of checks still involved differences in contour length, the stimuli were such that contour length could be assessed without the influence of number of checks. The results indicated that the number of checks and not contour length alone determined attention. The infants looked longer at the patterns having 32 checks. Fantz (1965) reports a study using stimuli that varied in the irregularity of arrangement of elements (squares), controlling for contour and area. Infants showed some tendency to look longer at the more irregular patterns and this was evident only after 2 to 3 months of age. These data suggest that either "complexity" apart from contour length is still a functional variable and/or that the perceptual system is influenced by contour length relative to the general size of the whole pattern and not to the absolute value of contour length.

*Summary*

The above data provoke the following analysis and set of hypotheses. The term "complexity" is too general, and the search for specific physical attributes of stimulus patterns, though infested with methodological problems, should be pursued more vigorously. Early data indicate that for visual stimuli the length of black-white contour is one of several aspects of visual arrays that governs attention. Contour length may operate by introducing a certain rate of change in retinal stimulation as the subject scans a pattern. If this is true, one should not expect the absolute length of contour to be as important as the length of contour relative to the total span of the stimulus. If an infant scans a pattern, he will do so only within the general expanse of the stimulus, and thus the amount of contour relative to the visual field defined by the stimulus will govern the number of black-white changes encountered during scanning and thus the amount of attention that the stimulus receives.

Further, it appears that as the infant matures he attends maximally to stimuli possessing more and more contour. For this kind of stimulation, age changes may level off (at least temporarily) between 5 and 9 months of age. It is tempting to speculate that the infant develops some type of sensory-perceptual coding rules during the early months of life which increase his efficiency in processing this kind of perceptual information. However, by the time the infant is between 5 and 9 months of age these early processes are well formed and further age changes in the distribution of attention await the development of other capabilities and different cognitive orientations toward this type of stimulus.

However, even given the above postulations, we still need to know (*a*) what other physical parameters of stimuli besides contour length are functional in determining attention, (*b*) what happens to the distribution of attention over a greater age range than we have currently studied, (*c*) what is the specific mechanism that mediates such age changes, and (*d*) what experiences can be given to infants to maximally facilitate such development.

## RESPONSE TO DISCREPANCY

The lure of studying the response of the infant to familiar and novel (or "discrepant") stimuli resides in the potential of the data in this area to tell us something of the familiarization process itself. One way to discern whether an infant remembers a given stimulus is to present him with a discrepancy, a stimulus that is similar yet distinctly different from the familiar stimulus. Suppose without any familiarization experience, two stimuli are equal in their power to recruit attention. If familiarization is given with one stimulus and if the infant responds differentially to the two stimuli as a function of one being familiar and the other novel or discrepant, it may be concluded that the infant remembers the familiarized stimulus (i.e., has some memory engram or internal referent for it) and discriminates the other as being different.[3] Such a rationale has been applied in a general way to studying object permanency (Charlesworth, 1966, 1968) and visual acuity (Fantz, Ordy, & Udelf, 1962). However, before the procedures may be used with confidence in the study of observational learning in infants, a great deal more must be known about the general phenomenon of attention to familiar vs. discrepant stimuli. For example, of

[3] If there is no difference in attention to the two stimuli, one cannot conclude that the infant did not discriminate between them because the learning-performance distinction must be maintained.

what significance is the role of the magnitude of discrepancy? How does the degree of familiarization affect the relative response to familiar vs. discrepant stimuli? Under what age and stimulus conditions is it methodologically feasible to make these assessments? For convenience in discussing these issues, accept the term "standard" to represent the stimulus to be made familiar, "memory engram" or "internal referent" for the internal perceptual-cognitive representation of the standard, and "discrepant stimulus" for the new stimulus presented to the infant.

*Age of Onset*

Until recently, the onset of attentional preference for discrepancy was placed at approximately 8 to 12 weeks of age. Curcio (1969), Fantz (1964), Uzgiris and Hunt (1965), and Weizmann, Cohen, and Pratt (1970) all present data which suggest that prior to 8 weeks of age the infant either does not differentiate or he looks more at the familiar standard in a two-choice situation. Sometime between 8 and 12 weeks of age, this propensity shifts and he begins to look longer at the discrepant member of the test pair. Each of these studies used a test procedure in which the standard and the discrepant stimuli were simultaneously present. It is likely that if a single stimulus presentation technique were used, requiring that the infant compare the new stimulus with a remembered engram of the familiar standard, the transition point would occur at an older age.

Interestingly, Siqueland (1969) has demonstrated recently that infants as young as one month of age will display greater attention to a novel vs. a familiar stimulus in a manner comparable to the performance of 4-month old infants. His procedure was to make the presentation (or withdrawal) of a stimulus contingent upon high amplitude sucking. After 5 minutes of responding for the same reinforcing stimulus, the color of that pattern was changed and a subsequent increment in sucking rate occurred. Though there appeared to be differences in general sucking rate between age groups, the response to the switch in reinforcers (i.e., to the introduction of a new stimulus) was comparable across age. Moreover, the sucking rate following the stimulus change was generally higher than throughout the entire initial conditioning phase, suggesting that the introduction of a new stimulus did more than merely dishabituate the infant and allow a return to response levels achieved prior to adaptation. It would appear, both on the basis of these data and on logical grounds, that very young infants respond to changes in stimulation but that the manifestation of this sensitivity is a function of the suitability of the assessment technique for the particular age being studied.

*Parameters of Discrepancy*

Are the data concordant with the proposition that infants consistently attend to the discrepant stimulus more than to the familiar standard even when the methodology is appropriate to reveal such tendencies? Although the general pattern of results favors such a positive conclusion, the effect has been qualified by the sex of the subjects (Meyers and Cantor, 1967), particular stimulus preferences (Saayman, Ames, & Moffett, 1964), and general type of stimulation (Lewis, Bartels, Fadel, & Campbell, 1966a; Meyers & Cantor, 1966).

Two factors have been investigated as possible variables underlying the inconsistency of results. First, it may be that some infants under certain conditions do not form an engram of the standard stimulus during the familiarization phase, and thus there would be no basis for attending differentially to "familiar" and "discrepant" stimuli. One approach to studying this possibility is to use long-term familiarization techniques in which the infant is presented with the standard over several weeks. Presumably, this should allow sufficient time for an engram to be formed. Another route is to assume that the rate of habituation of attention to the repeated presentation of the standard reflects the acquisition of a memory engram for that standard. Then, infants who demonstrate such habituation should respond positively to a subsequent discrepancy whereas infants who fail to habituate might not.

A second factor potentially involved in the inconsistency of findings is that the amount of attention paid to a discrepant stimulus might be a function of the magnitude of discrepancy which that stimulus represents from some well-familiarized standard. This possibility derives from the "discrepancy hypothesis" which has been proposed in a variety of other contexts (e.g., Dember & Earl, 1957; Hebb, 1949; Hunt, 1963, 1965; McCall, 1965, 1970; McClelland, Atkinson, Clark, & Lowell, 1953). Applied to the present situation, attention is predicted to be a curvilinear function of the magnitude of discrepancy from the standard as displayed in Figure 2. If this is true, consider the implications of the traditional procedure of assessing attention to just two stimuli, the standard and one discrepant stimulus. If the new stimulus represents *A* units of discrepancy (see Figure 2), the theory predicts that it should receive more attention than the standard. However, if the discrepant stimulus represents *B* or *C* units of discrepancy, the attentional response to it may be equal to or even less than that accorded to the familiar standard. To the extent that the discrepancy hypothesis is valid, dichotomous assessment of familiar–discrepant stimuli will yield ambiguous results.

The following sections consider the tenability of these speculations on the length of familiarization, habituation, and magnitude of discrepancy as variables in the distribution of attention.

*Amount of familiarization.* Intuitively, the amount of familiarization must be a determinant of the differential response to familiar and discrepant stimuli. McCall and Melson (1970b), gave 5½-month old males either 4, 8, or 12 presentations of an 8-note tonal sequence and then presented a discrepancy formed by rearranging the original eight notes into a new sequence. The

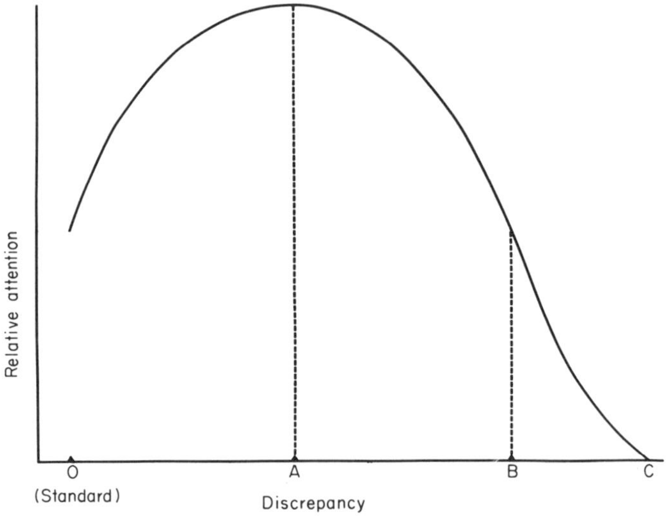


**Fig. 2.** One theoretical relationship between the magnitude of discrepancy and relative attention.

differential attention (i.e., cardiac deceleration) to the discrepant sequence relative to the preceding standard was a linear increasing function of the magnitude of familiarity with the standard. Analysis revealed that this was not simply a function of greater habituation to the standard by those infants given 12 presentations of it. In fact, most of the habituation was completed by the end of four presentations.

The above experiment suggests that differences in familiarization within the range of traditional procedures can moderate the response to discrepancies. If the differential ability to acquire a memory engram of a stimulus is the major underlying variable contributing to the inconsistencies in results cited above, then giving infants an opportunity to become familiar with the standard over several weeks should guarantee that all infants will develop an internal referent for the standard, and the differential attention to familiar and discrepant stimuli should be less equivocal.

*Long-term familiarization.* As cited above, some research indicates that 2 to 3 months of age is a possible transition point after which the infant will allocate more attention to a discrepancy than to the standard if a paired comparison procedure is used. If the stimuli were presented one at a time (because cardiac deceleration was to be used as a dependent measure), it was supposed that infants would be able to perform this more difficult task by 4 months of age.

Consequently, McCall and Kagan (1967a) gave mothers of 3-month old infants a standard stimulus in the form of a mobile (see Figure 3) to take home to present to their infants several times per week for a period of 1 month. Then, the familiarized infants and a group of nonexperienced controls were shown that standard and three graded discrepancies from it. The standard was considered to be a vertical linear array of elements while the discrepancies deviated from the standard first in verticality and then also in linearity. Adults who rated the similarity of the discrepancies to the standard also scaled the stimuli in this order.[4]

In terms of cardiac deceleration, the experienced girls ("Girls-Mobile") displayed the expected curve as shown in Figure 4, increasing attention with increasing amounts of discrepancy, up to a point. The nonfamiliarized control group evidenced no differential response to the four stimuli. Notice that the response to the two and three-unit discrepancy was not merely a dishabituation phenomenon since the cardiac deceleration to these stimuli was higher than for the control group for whom these stimuli were totally novel. A stimulus made discrepant by familiarizing the infant with a similar but different stimulus apparently possesses greater power to recruit attention than if that same stimulus were totally novel.

However, as is clear from Figure 4, the experienced boys ("Boys-Mobile") did not differ from controls. This sex difference was unexpected. Perhaps 4 months of age is very near the lower age limit for infants to be able to perform such a task. Since girls are more precocious in a variety of other areas, it is possible that the girls, but not the boys, were sufficiently advanced to perform this task. Alternatively, a static,[5] unresponsive stimulus may be less capable of recruiting the involved attention of male than female infants (see later section on sex differences).

Another hypothesis is provoked by a study reported by Weizmann *et al.* 1971), who manipulated the familiarity of the infant with both the test stimulus and with the test surround (the bassinet). They found that boys responded differentially to the discrepant and familiar stimuli when the surround was familiar but not when it was novel. The results were just the opposite for girls. These data suggest that perhaps the boys are more attentive to the test environment which distracts them from the experimental stimuli. This general phenomenon (though not necessarily the sex difference) has been observed with rats (McCall, Weiffenbach, & Tucker, 1967; Sheldon, 1969).

[4] It is acknowledged that because adults perceive stimuli to fall on a given scale this does not guarantee that infants also perceive them in the same way. The use of a theory (e.g., information theory) to scale stimuli does not alleviate one problem. Thus, there is no easy solution to scaling stimuli for infants, and one must acknowledge the assumptions being made and cautiously proceed.

[5] Figure 4 describes the experimental group as having experienced "mobiles." This is a misnomer since the stimuli did not move very much and were not unde; the control of the infant. "Stabile" would have been more descriptively accurate.

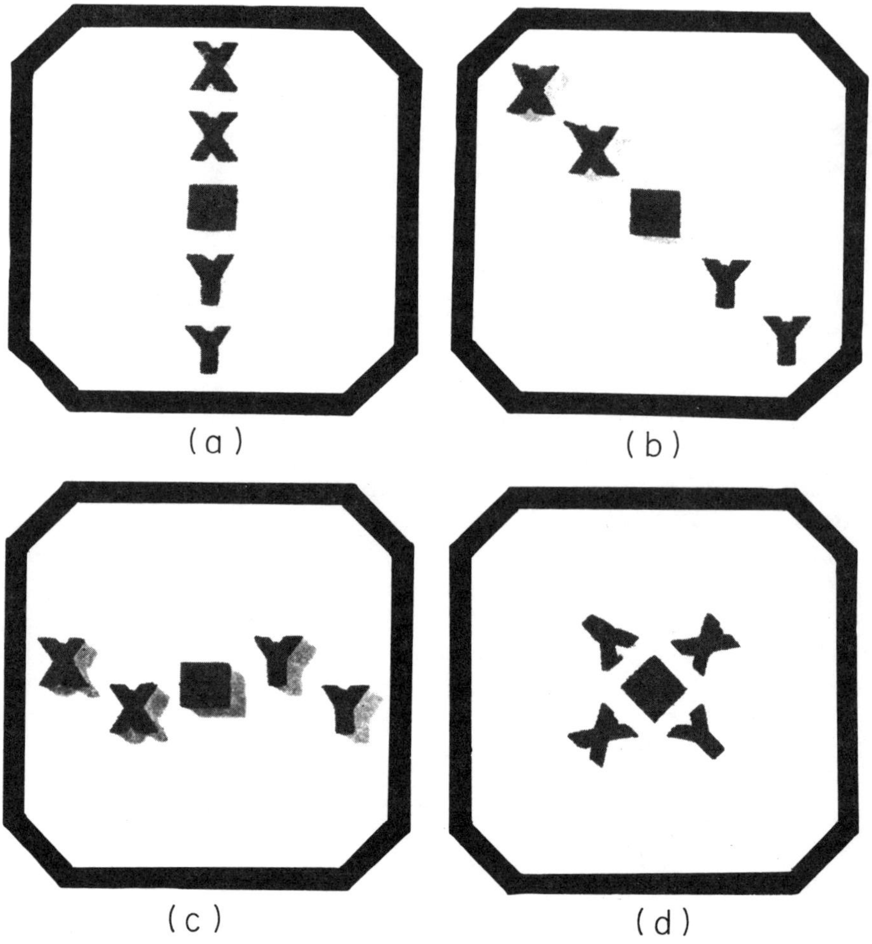


**Fig. 3.** Stimuli used by McCall and Kagan (1967a) and McCall and Melson (1969). Stimulus *a* was the standard and Stimuli *b*, *c*, and *d* were conceived to be increasing levels of discrepancy from *a*. (From McCall & Melson, 1969; reproduced by permission of authors and publisher.)

In short, the long-term familiarization paradigm seems to yield results consonant with the discrepancy hypothesis for girls. Some factor relevant to the saliency of the stimuli or the surround appears to interfere with the phenomenon for boys. Whatever this factor is, merely providing ample time for an engram to be developed is not a sufficient condition for observing an attentional preference for discrepant stimuli among boys within the context of long-term familiarization studies.

*Short-term familiarization.* It is possible that some infants, for one reason or

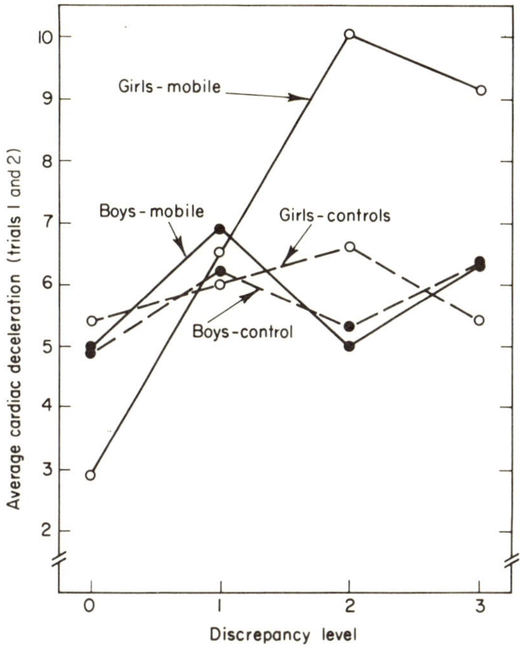


**Fig. 4.** Mean cardiac deceleration for the four stimuli plotted separately for sexes and for familiarized ("mobile") or control groups. (From McCall & Kagan, 1967a; reproduced by permission of authors and publisher.)

another, do not form engrams of the standard stimulus. If it is assumed that the rate of habituation of attention to a repeatedly presented standard is an index of the acquisition of an engram for that stimulus, then infants that display rapid habituation should respond more positively to discrepancies than those subjects who do not evidence habituation. Consequently, a series of studies was devoted toward examining the tenability of the discrepancy hypothesis by giving infants familiarization with a standard within a single experimental session, and toward examining the rate of habituation during the familiarization process as a predictor of the response to discrepancy.

The purpose of an initial experiment (McCall & Kagan, 1970) was to ask whether 4-month old infants would respond to a discrepant stimulus after only five trials of familiarization with the standard, and whether the pattern of response would vary with the rate of habituation during the familiarization period and with the magnitude of discrepancy.

Subjects were 36 boys and 36 girls, 4 months of age. Each stimulus was composed of three elements (e.g., a blue and green flower, a red and silver Christmas bow, a small plastic Scottish terrier, etc.) arranged in a triangular pattern. The magnitude of discrepancy was formed by replacing one, two, or all three of these stimulus elements with other qualitatively different elements. Each infant saw the following stimulus sequence:

SSSSSCSSSCSSSCSSS

in which S is a standard stimulus (the same particular stimulus on each presentation) and C is a changed stimulus representing 1, 2, or 3 arbitrary units of discrepancy (not the same magnitude on each presentation of a C). The study was counterbalanced for particular stimuli and stimulus elements.

Originally it was intended to divide the subjects into two groups, those showing either rapid or slow habituation of first fixation during the familiarization phase of the procedure. However, it was found that some children did not look very long from the beginning of the session, and therefore it was not meaningful to consider these children as rapid or slow habituators since their looking times were consistently short and could not show much decline over trials. Therefore, three "habituation" groups were formed: *short lookers, rapid habituators,* and *slow habituators.*

The results are presented in Figure 5. The magnitude of discrepancy has not been considered here; only the general response to a change in stimulus is indicated. The graph reveals that the short lookers responded positively to each presentation of a discrepancy, the rapid habituators did so only to the first discrepancy, and the slow habituators showed only minimal response to any of the three discrepancies.[6]

Since the groups were classified on the basis of first fixation and the results in Figure 5 are also in terms of first fixation, an attempt was made to assess the response to discrepancies in terms of another measure, specifically smiling. First, consider the number of infants who smiled to the discrepancies (regardless of whether they also smiled to the standard): 27% of the short lookers, 19% of the rapid habituators, but none of the slow habituators smiled to a discrepant stimulus. Further, there were four infants in the study who smiled only to a discrepancy and never to the standard (disregarding the first five standards); all of these were short lookers. These smiling data tend to support the proposition that the pattern of first fixation displayed during the habituation phase of the experiment predicts the responsivity of infants to a discrepancy.

Since only the rapid habituators and short lookers responded positively to a discrepancy (ignoring magnitude), the analysis of the effect of magnitude of discrepancy was limited to those subjects. First fixation to the discrepancies relative to the standard was found to be an increasing linear function of the magnitude of discrepancy. Interestingly, there were no major sex differences observed in this study; boys as well as girls responded to short-term discrepancy.

[6] Actually, since the first stimulus was not considered in forming the habituation groups and since the short-lookers did in fact habituate from trial 1 to 2, the short-lookers may have been the "most rapid habituators." In this case, the results indicate a direct relation between habituation rate and response to discrepancy.

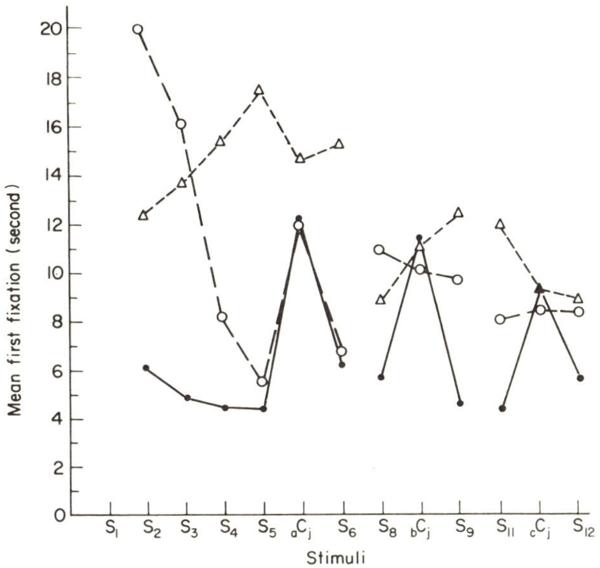


**Fig. 5.** Mean first fixation for the three habituation groups. ●—● Short look (N = 26); ○– –○ rapid habit (N = 16); △– –△ slow habit (N = 18). (From McCall & Kagan, 1970. Copyright (1970) by the American Psychological Association; reproduced by permission.)

Thus, when stimuli are presented a few short times, the sex difference observed under conditions of long-term familiarity was not found.

These data suggested that the pattern of habituation and the magnitude of discrepancy were potential variables in the deployment of attention. While encouraging, replication and extension of these results were necessary. Consequently, a second experiment (Melson & McCall, 1970) was performed to determine whether infants would respond with cardiac deceleration to discrepancies defined by alterations in tonal sequences and whether or not the magnitude of discrepancy and the rate of cardiac habituation during the familiarization phase would predict the response to auditory discrepancies.

Because of other circumstances, the sample was restricted to girls, and 48 five-month infants were subjects. One stimulus was a simple C-major ascending scale of eight notes. A second stimulus was the same except that the positions of the first and sixth notes were reversed. The difference between these two tonal sequences defined the small discrepancy. A third stimulus was composed of the same eight notes from the simple ascending scale but their sequence was rearranged so that no note occupied the same position or followed the same note as in the ascending scale. The difference between the ascending scale and this third stimulus constituted the large discrepancy condition. Within each of the discrepancy groups, some subjects received the ascending series as their standard while others had the nonascending pattern. Any single subject listened to only two different stimuli, a standard (S) and a discrepancy (D) in the following sequence:

SSSSSSSSDSSDSSDSSDS.

In addition, subjects were independently divided into rapid and slow habituators as a function of the extent to which they showed progressive declines in cardiac deceleration to the eight presentations of the standard prior to the first discrepancy.

In terms of cardiac deceleration, the results indicated that rapid habituators responded more than slow habituators to a discrepancy relative to the preceding standard. It was unlikely that this was simply a function of initial values because there were no differences between habituation groups in absolute heart rate just prior to the first discrepancy. Further, as seen in Figure 6, rapid habituators tended to respond consistently to a large discrepancy whereas they did not respond positively to a relatively smaller discrepancy after the second presentation. The slow habituators, depicted at the top of Figure 6, did not evidence such trends. In short, rapid habituators responded more to auditory discrepancies than did slow habituators, and among rapid habituators there was a tendency to perseverate longer in the response to large than to small discrepancies.

It happened that all of these subjects had experienced a visual series prior to the auditory episode described above (McCall & Melson, 1970a). Since subjects saw four stimuli and not just one in the visual series, it was not appropriate to classify subjects according to their rate of habituation to visual stimuli. However, it was possible to group subjects into short vs. long lookers, a variable that was found to be salient in the McCall and Kagan (1970) experiment described above. The question was whether results could be obtained using auditory stimuli and a cardiac measure of attention that parallelled those previously obtained for visual stimuli and first fixation.

There was a tendency for short lookers to respond more than long lookers to auditory discrepancies relative to previously presented standard stimuli. It was also observed that short lookers smiled more discriminatively (*either* to the familiar *or* to the discrepancy but not to both) than long lookers, and the three infants who smiled to the discrepancy and not to the standard were all short lookers. These results conform nicely to those found previously by McCall and Kagan (1970) for visual stimuli and suggest that the original results may generalize across stimulus modality as well as response measure (first fixation and cardiac deceleration).

It has been suggested above that both length of looking at a visual stimulus and rate of cardiac habituation to a repeatedly presented auditory stimulus predict the response to auditory discrepancies in terms of cardiac deceleration. Interestingly, there was no relationship between these two predictors (an $r$ of .03 between first fixation time to visual stimuli and a single index of rate of cardiac habituation to auditory stimuli). Furthermore, among rapid habituators,

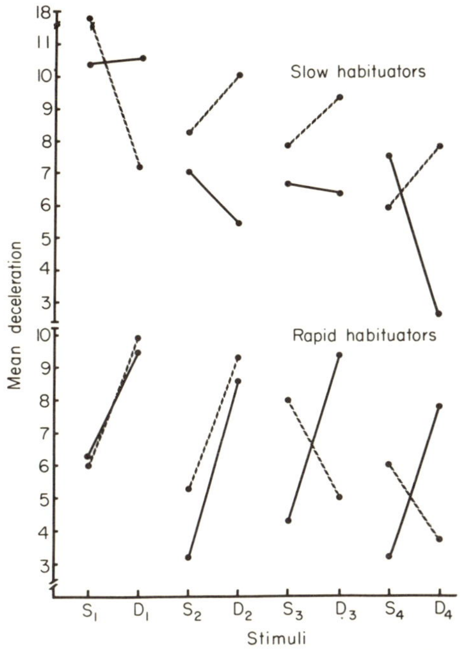


**Fig. 6.** Mean cardiac deceleration for slow and rapid habituators as a function of magnitude of discrepancy and trial. - - - small disc; —— large disc. (From Melson & McCall, 1970; reproduced by permission of authors and publisher.)

short lookers respond more to auditory discrepancies than long lookers. Among slow habituators, the same was true. In fact, slow habituating, long looking subjects responded less to the discrepant stimulus than to the standard. These data suggest the hypothesis that short and long looking to visual stimuli and cardiac habituation to auditory stimuli are independent and summative predictors of the response to auditory discrepancy. Perhaps a general propensity towards habituation exists, but one that is also keyed separately to the different modalities.

A third attempt (McCall & Melson, 1969) was made to demonstrate that magnitude of discrepancy and habituation rate are variables that determine the amount of attention allotted to a stimulus.

Twenty-one males, 5½ months of age, were divided into three groups which experienced different magnitudes of visual discrepancy. The stimuli were those displayed in Figure 3. The standard stimulus for all subjects was the vertical linear array of five stimulus elements (X's, Y's, and a square). From the previous data (McCall & Kagan, 1967a), it was assumed that the infants had no *a priori* attentional preferences among these stimuli. All subjects saw the following stimulus sequence:

SSSSSSSSDSSDSSDS

The results are presented in Figure 7. The experiment was designed to test for magnitude of discrepancy and such an effect is illustrated by the solid line. Small magnitudes of discrepancy were responded to with more cardiac deceleration than previously presented standards while the largest magnitude of discrepancy was actually attended to less than the standard. The dotted line connects this pattern to the point representing the amount of attention paid to the standard. Since a difference score was used as the dependent measure, zero represents an equivalent response to discrepancy and standard. Although previous results indicated only the increasing portion of the hypothesized curve (Figure 2), these data hint at the curvilinear function proposed by the discrepancy hypothesis. Thus, a discrepancy received more, the same, or less attention than the standard as a function of the magnitude of discrepancy.

Since each subject in this observation viewed the standard eight times prior to the first discrepancy, the rate of habituation as well as the response to discrepancy could be computed both in terms of first fixation and cardiac deceleration. After treatment effects were statistically removed, indices of habituation and of response to discrepancy were correlated. Habituation to visual stimuli in terms of cardiac deceleration predicted neither the response to visual discrepancy in terms of first fixation nor in terms of cardiac deceleration. However, habituation to visual stimuli in terms of first fixation predicted the response to discrepancies both in terms of first fixation ($r = .75$) and in terms of cardiac deceleration ($r = .52$). It would appear that habituation to visual stimuli in terms of first fixation (but not cardiac deceleration) predicts the response to discrepancy in terms of either of these two response measures. It is also important to note the fact that visual habituation predicted the cardiac response to discrepancy tends to minimize claims that such effects are artifacts of initial values. However, it is true that first fixation and cardiac deceleration are not totally independent measures of attention, and to this extent the argument against initial values is not totally conclusive. In summary, these findings again show that habituation predicts the response to discrepancy, but only when habituation is measured in terms of first fixation. It may be that habituation of first fixation is the salient variable for visual stimuli while habituation of cardiac deceleration is the appropriate measure for auditory stimuli.

*Summary*

The substance of these results is that in the context of short-term familiarization the rate of habituation of attention to the repeated presentation of a stimulus predicts the response to discrepancies in terms of a variety of

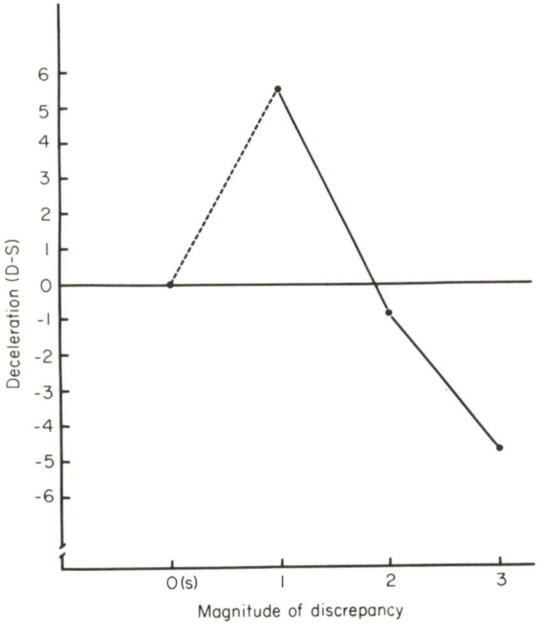


**Fig. 7.** Cardiac deceleration to the discrepancy relative to the preceding familiar standard as a function of the magnitude of the discrepancy. (From McCall & Melson, 1969; reproduced by permission of authors and publisher.)

behaviors. Infants who display rapid habituation are more likely to respond with increments in attention to discrepancies relative to the standard than are infants who do not indicate rapid habituation. Further, there is some evidence to suggest that the pattern of habituation (technically long vs. short looking) to a visual stimulus may predict the response to auditory discrepancies as measured in a separate episode with another response variable (i.e., cardiac deceleration). Finally, cardiac habituation to auditory stimuli but visual habituation to visual stimuli may represent the salient predictor variables within these two modalities, respectively.

A second general result is that given sufficient habituation to the standard, the magnitude of discrepancy determines the amount of attention a new stimulus will receive. Although the discrepancy hypothesis predicts an inverted-U relationship between magnitude of discrepancy and relative attention, the theory does not prescribe how to select a stimulus dimension that will reveal the entire function. Consequently, when attention is found to be an increasing function of discrepancy, one cannot decide whether the prediction of a curvilinear function is in error or whether the stimulus dimension used in the experiment did not include stimuli of sufficient discrepancy to obtain the complete curve. Some data cited here and results in other fields of study (see McCall, 1970) suggest that the inverted-U does occur under some conditions, but these circumstances cannot yet be specified.

If one assumes that an increase in discrepancy is associated with an increase in cognitive uncertainty, then these data promote the hypothesis that the infant attends to that which provides him with a moderate amount of new information—enough information to require some time to process—but not information for which the requisite cognitive structures to process and assimilate it are not available. An interesting speculation is that this "optimal amount of information" also represents a level of input that can most efficiently affect accommodative changes in the cognitive structure itself (Dember & Earl, 1957; Hunt, 1963, 1965; McCall, 1970; Munsinger & Kessen, 1964, 1966; Piaget, 1952.

*The Meaning of Habituation*

Throughout this discussion it has been assumed that habituation of the attentional response to the repeated presentation of a stimulus reflects the acquisition of a memory engram for that stimulus. Lewis (1967) has reviewed and presented several sources of data that bolster the general construct validity of habituation for this purpose.

1. Habituation rate predicts the response to discrepancy.
2. Younger infants display less habituation over trials than older subjects (Curcio, 1969; Fantz, 1964; Fantz & Nevis, 1967; Lewis, 1967; Lewis *et al.*, 1966a; Lewis, Goldberg, & Rausch, 1966b).
3. Young infants require a shorter interstimulus interval to demonstrate habituation than do older subjects (Lewis, 1967).
4. Habituation is not likely to be a simple function of fatigue or general adaptation to the experimental situation (Cohen, 1969a).
5. Brain damage and other maladies (e.g., low Apgar score) may inhibit habituation (e.g., Lewis, 1967; Lewis, Bartels, Campbell, & Goldberg, 1967; Thompson & Spencer, 1966).
6. Habituation is slower to more complex stimuli than to relatively simple patterns (Caron & Caron, 1968).
7. Perceptual habituation at one year of age correlates with Stanford-Binet IQ and performance on a concept formation task at 44 months of age (Lewis, 1967), but Kagan (personal communication) has not found this using somewhat different procedures .

These data are consistent with the proposition that habituation reflects a cognitive process, perhaps the acquisition of a perceptual memory. However, as customarily measured, habituation appears to be neither a necessary nor a sufficient condition for the occurrence of an increment in response to a discrepancy. There have been reports of habituation without a response to discrepancy (Lewis *et al.*, 1966a), no apparent habituation but a positive response to a discrepancy (McCall & Kagan, 1970; Meyers & Cantor, 1967), and increased response to discrepancy with increasing numbers of familiarization trials but without corresponding differences in level of habituation (McCall & Melson, 1970b).

What, then, does response habituation mean? How can one explain the fact that some infants who habituate do not respond to discrepancies? One possibility is that some infants just "tune out" and become bored with the entire experimental situation. They would "habituate" but not respond to a discrepancy. A more threatening possibility is that engram acquisition is displayed in the habituation of one response measure while the attentional response to a discrepancy is reflected in another behavioral index, and the experimenter may or may not be measuring one or both of these variables. For example, an infant girl evidenced a beautiful pattern of cardiac habituation to an auditory stimulus. When observers returned to the polygraph room after the discrepancy had been presented they exclaimed how the discrepant stimulus evoked a facial expression of surprise and a widening of the eyes. In spite of this report, there was clearly no response to the discrepancy in terms of cardiac change, the variable being measured. The observation that habituation of one variable (first fixation) but not another (cardiac deceleration) can predict the response to discrepancy in terms of many dependent variables (fixation, deceleration, smiling) makes this general point more plausible. Further, McCall and Melson (1970b) have observed an inverse relationship (−.70) between the relative cardiac responsiveness of male infants (females not tested) to auditory and to visual discrepancies, suggesting that some infants may be more tuned to changes in stimulation in one than in another modality. Collectively, these data point to our need for a theory of the dynamics of attentional behavior (i.e., the response measures), since it appears likely that not only are there interactions between response measures that reflect stimulus events, on the one hand, and the modality of stimulation, on the other, but that these two variables in turn interact with individual differences among subjects. It is quite possible that the attentional response that reflects engram acquisition or response to discrepancy is more idiosyncratic than we care to imagine.

With respect to infants who do not show response habituation but do respond to discrepancies, it may be that some subjects habituate very rapidly—perhaps, within a single stimulus presentation. Thus, no response decline over trials would be apparent. Alternatively, it is possible that an infant who continues to observe a stimulus over several trials without habituation is really giving it "a very thorough study." Such nonhabituating subjects would respond to discrepancies; in fact, they may even have a stronger engram for the standard than rapidly habituating infants. In short, habituation rate seems to predict the response to discrepancies, but the more theoretical question of what response habituation means is still somewhat ambiguous.

*Sex Differences*

Sex differences in the deployment of attention in infants abound. These effects are difficult to evaluate for several reasons, one of which is that some investigators do not even report how many subjects of each sex were used while others do not analyze their data for differences between the sexes. Further. the pattern of results that have been reported is not particularly uniform or clear, and long-term familiarization studies yield different results than short-term procedures. What follows is a brief summary of some of the sex differences that have been observed, organized around two highly speculative propositions which are offered mainly for heuristic purposes.

The first proposition is that males are more cued or dependent upon a responsive stimulus than are females. Males are less likely to attend to and learn from long-term exposure to physically static stimuli than are females. McCall and Kagan (1967a) reported that after exposure to a fixed visual stimulus for 1 month, girls but not boys paid differential attention to discrepant stimuli. Data from experiments that familiarize within a single session do not support this sex difference. Presumably, short-term procedures are successful in capturing the attention of both males and females for at least a short period of time. In another line of evidence, Kagan (personal communication) has reported Kaye's finding that the tendency of infants to reach out and touch a new stimulus dropped out for girls in middle infancy but remained stable for boys, suggesting that the desire to manipulate an object retains its strength over age for boys but not girls.

The manipulative exploration data of McCall and Garratt (1969), showing that boys are more responsive to differences in the sound potential and manipulability of objects than are girls, fits nicely with this proposition. Goldberg and Lewis (1969) observed that 13-month old boys were more active and vigorous in their manipulative play while girls concentrated their activities on fine as opposed to gross motor skills. Thus, these data are at least consistent with the proposition that boys are more geared to a motor involvement with a responsive stimulus than are girls, though it may be that they just take longer to grow out of this early stage. One might contemplate the qualifications such a proposition makes for theories that emphasize a sensorimotor interaction as the basis of development (e.g., Held, 1965; Piaget, 1952).

A second proposition is that in the context of short-term familiarization there is a slight difference in modality preference for boys and girls. Auditory stimuli and vocal behavior appear to be more salient for girls than for boys, and the visual mode may be more compelling under some circumstances for males than for females. Watson (1969) attempted to condition ocular and head orientation with visual or auditory reinforcement in 10- and 14-week old infants. At 10 weeks of age, girls learned with auditory and with auditory-plus-visual reinforcement while boys did not learn under any condition. At 14 weeks of age, girls learned with auditory but not visual reward while boys learned with visual but not auditory stimulation. Pancratz and Cohen (1970) found that boys habituated (first fixation) to a green circle more than girls and also showed more response to a change in that stimulus than did girls. Conversely, Meyers and Cantor (1966) state that girls habituated more (cardiac deceleration) than boys to visual stimuli, but the variety of stimuli presented to the infants increased with the number of trials. Thus, failure to habituate in this context may reveal a greater interest by the boys in the newly introduced and changing set of stimuli. Horowitz and Paden (1969) also report that girls habituate more rapidly than boys to visual stimuli, but they present no group data or statistical evaluation of this effect. Meyers and Cantor (1967) observed that boys but not girls gave differential attention (deceleration) to the several visual stimuli in their experiment. Thus, the data on differential habituation to visual stimuli is partially in support of the hypothesis and two experiments show boys giving greater response to visual discrepancies than girls. It must be acknowledged that similar studies report no major sex differences (e.g., McCall & Kagan, 1970), so that much work needs to be done to specify the conditions under which sex differences will be found.

Another line of research tends to support the proposition that auditory stimuli and vocal behavior are more salient for the female. Two studies, as well as unpublished data from our own laboratory, have now shown that early vocalization on standardized infant tests predicts later verbal IQ for girls, but not for boys (Cameron, Livson, & Bayley, 1967; Moore, 1967). Kagan (1968) has demonstrated that vocalization represents a stable individual difference for girls throughout infancy but not for boys.

Though highly speculative, it appears that a reasonable working hypothesis concerning sex differences is that boys are more cued to a motoric and responsive interaction with objects whereas girls are more able to profit from interaction with a static stimulus than are boys. Further, auditory stimuli and vocal behavior may be more salient for girls; boys may be more tuned to visual events under some circumstances. The data are tentatively suggestive when sex differences have been observed, but as yet one is unable to predict the precise circumstances (e.g., long-term vs. short-term familiarization) under which sex differences will be manifested.

## CONCLUSIONS

The study of the distribution of attention holds promise for understanding the cognitive development of infants. The data support the proposition that the infant's perceptual-cognitive system develops during the first several months of life and progressively he comes to give preferential attention to patterns of increasing "complexity," at least up to a point. It is likely that the salient component of visual stimuli varying in "complexity" is the length of black-white contour relative to the visual area of the pattern, which in turn may be subserviant to the more general variable of rate of change in stimulation. Just what the mechanisms are that develop during the first months and presumably allow the infant to process more and more intricate stimuli are not known.

Further, the habituation of attention with repeated presentations of a given stimulus may be useful as an index of the extent to which the infant has formed a memory engram of that stimulus. Habituation, then, might be used to determine which stimulus configurations are most easily learned by the infant. Regrettably, there are some data which indicate that habituation may not be either a necessary or a sufficient index of the acquisition of a memory for a stimulus, and it is not yet known whether rapid habituation means rapid learning or very limited learning about that stimulus.

Infants do respond differentially to familiar and discrepant stimuli, and this fact would seem to open the door to studying the perceptual-cognitive learning process itself as it is reflected in the relative attentional response to discrepancies. However, it appears that the amount of attention paid to a stimulus is a function of the magnitude of discrepancy of that stimulus from what the infant knows well. Further, if the entire range of discrepancy magnitudes is tested, that relationship may not be linear but an inverted-U function. It has been hypothesized that the stimulus event that recruits the maximum attention also has the greatest likelihood of affecting accommodative changes in the perceptual-cognitive structures themselves.

There is a certain intuitive appeal to the study of attention as an index of cognitive functioning. This feeling has been bolstered both by the data cited above and also by the fact that institutional infants do not show differential attention to novel stimuli as early as home-reared children (Fantz & Nevis, 1967) and that handling and enriching experiences during infancy facilitate the development of attentional processes (Ottinger, Blatchley, & Denenberg, 1968; Rubenstein, 1967; White, 1969). This has raised the hope among some that assessments of infant attentional behavior may be more salient predictors of adult intellectual performance than our current infant "intelligence" tests. Certainly, the studies reporting correlations (as high as the .70s) between infant vocalization behavior and adult (e.g., 26 years) verbal intelligence are encouraging (Cameron *et al.,* 1967; Moore, 1967). Further, Fantz and Nevis (1967) report that the differential attention to novel and familiar stimuli discriminated between institutional and home-reared infants at 2 to 3 months of age, while the Griffiths Mental Development Scale did not do so until 15 months of age.

Consequently, there is substantial hope attached to the study of attention as an index of cognitive functioning, but as usual the road has been rocky and winding. It is obvious that attention is a cluster concept and that its many response facets are not mutually equivalent. There is a great need for a theory that will specify the meaning of various responses and predict which response (e.g., fixation time, deceleration, etc.) will reveal predicted effects. Further, why do we observe sex differences in some contexts but not others and what are the dynamics that produce these differences? Can we predict which infants will respond with a cardiac change, which will show a greater sensitivity to discrepancies in one modality than another, etc? Despite these impeding questions and the complicated nature of the task, the pursuit of the understanding of cognitive development in infancy through the study of attentional processes remains promising and exciting.

## ACKNOWLEDGMENTS

The author thanks Virginia Crandall, J. Kagan, R. M. Liebert, and F. Weizmann for their helpful comments on early drafts of this paper and to Joanne Peterson for her help in preparing the manuscript.

## REFERENCES

Berlyne, D. E. The influence of the albedo and complexity of stimuli on visual fixation in the human infant. *British Journal of Psychology,* 1958, vol. 49, 315-318.

Brennan, W. N., Ames, E. W., & Moore, R. W. Age differences in infants' attention to patterns of different complexities. *Science,* 1966, vol. 151, 354-356.

Cameron, J., Livson, N., & Bayley, N. Infant vocalizations and their relationship to mature intelligence. *Science,* 1967, vol. 157, 331-333.

Caron, R. F., & Caron, A. J. The effects of repeated exposure and stimulus complexity on visual fixation in infants. *Psychonomic Science,* 1968, vol. 10, 207-208.

Charlesworth, W. R. Persistence of orienting and attending behavior in infants as a function of stimulus-locus uncertainty. *Child Development,* 1966, vol. 37, 473-491.

Charlesworth, W. R. Cognition in infancy: Where do we stand in the mid-sixties? *Merrill-Palmer Quarterly,* 1968, vol. 14, 25-46.

Cohen, L. B. Observing responses, visual preferences, and habituation to visual stimuli in infants. *Journal of Experimental Child Psychology,* 1969, vol. 7, 419-433. (a)

Cohen, L. B. Alternative measures of infant attention. Paper presented at the meeting of the Society for Research in Child Development, Santa Monica, California, March, 1969. (b)

Curcio, F. The role of age and experience in infants' selective attention to novelty and incongruity. Paper presented at the meeting of the Society for Research in Child Development, Santa Monica, California, March, 1969.

Dember, W. N., & Earl, R. W. Analysis of exploratory, manipulatory, and curiosity behaviors. *Psychological Review,* 1957, vol. 64, 91-96.

Eisenberg, R. B., Griffin, E. J., Coursin, D. B., & Hunter, M. A. Auditory behavior in the neonate. *Journal of Speech and Hearing Research,* 1964, vol. 7, 245-269.

Fantz, R. L. Pattern vision in young infants. *Psychological Record,* 1958, vol. 8, 43-47.

Fantz, R. L. The origin of form perception. *Scientific American,* 1961, vol. 204(5), 66-72.

Fantz, R. L. Pattern vision in newborn infants. *Science,* 1963, vol. 140, 296-297.

Fantz, R. L. Visual experience in infants: Decreased attention to familiar patterns relative to novel ones. *Science,* 1964, vol. 146, 668-670.

Fantz, R. L. Visual perception from birth as shown by pattern selectivity. *Annals of The New York Academy of Sciences,* 1965, vol. 118, 793-814.

Fantz, R. L., & Nevis, S. The predictive value of changes in visual preferences in early infancy. In J. Hellmuth (Ed.), *The exceptional infant,* Vol. 1. Seattle: Special Child Publications, 1967.

Fantz, R. L., Ordy, J. M., & Udelf, M. S. Maturation of pattern vision in infants during the first six months. *Journal of Comparative and Physiological Psychology,* 1962, vol. 55, 907-917.

Fitzgerald, H. E. Autonomic pupillary reflex activity during early infancy and its relation to social and nonsocial visual stimuli. *Journal of Experimental Child Psychology,* 1968, vol. 6, 470-482.

Goldberg, S. & Lewis, M. Play behavior in the year-old infant: Early sex differences. *Child Development,* 1969, vol. 40, 21-32.

Graham, F. K., & Clifton, R. K. Heart-rate change as a component of the orienting response. *Psychological Bulletin,* 1966, vol. 65, 305-320.

Haith, M. M. The response of the human newborn to visual movement. *Journal of Experimental Child Psychology,* 1966, vol. 3, 235-243.

Haith, M. M., Kessen, W., & Collins, D. Response of the human infant to level of complexity of intermittent visual movement. *Journal of Experimental Child Psychology,* 1969, vol. 7, 52-69.

Hebb, D. O. *The organization of behavior.* New York: Wiley, 1949.

Held, R. Plasticity in sensory-motor systems. *Scientific American,* 1965, vol. 213, 84-94.

Hershenson, M. Visual discrimination in the human newborn. *Journal of Comparative and Physiological Psychology,* 1964, vol. 58, 270-276.

Hershenson, M., Munsinger, H., & Kessen, W. Preference for shapes of intermediate variability in the newborn human. *Science,* 1965, vol. 147, 630-631.

Horowitz, F. D., & Paden, L. Y. Infant control of visual stimulation three to fourteen weeks of age. Paper presented at the meeting of the Society for Research in Child Development, Santa Monica, California, March, 1969.

Hunt, J. McV. *Intelligence and experience.* New York: Ronald Press, 1961.

Hunt, J. McV. Motivation inherent in information processing and action. In O. J. Harvey (Ed.), *Motivation and social interaction: Cognitive determinants.* New York: Ronald Press, 1963. Pp. 35-94.

Hunt, J. McV. Intrinsic motivation and its role in psychological development. In D. Levine (Ed.), *Nebraska symposium on motivation.* Lincoln: University of Nebraska Press, 1965. Pp. 189-282.

Jeffrey, W. E. The orienting reflex and attention in cognitive development. *Psychological Review,* 1968, vol. 75, 323-334.

Kagan, J. The distribution of attention in infancy. Paper presented at the Association for Research in Nervous and Mental Disease, New York, 1968.

Kagan, J., Henker, B. A., Hen-Tov, A., Levine, J., & Lewis, M. Infants' differential reactions to familiar and distorted faces. *Child Development,* 1966, vol. 37, 519-532.

Kagan, J., & Lewis, M. Studies of attention in the human infant. *Merrill-Palmer Quarterly,* 1965, vol. 11, 95-127.

Karmel, B. Z. The effect of age, complexity, and amount of contour on pattern preferences in human infants. *Journal of Experimental Child Psychology,* 1969, vol. 7, 339-354.

Lacey, J. I. Somatic response patterning and stress: Some revisions of activation theory. In M. H. Appley & R. Trumbull (Eds.), *Psychological stress: Issues in research.* New York: Appleton-Century-Crofts, 1967.

Lacey, J. I., Kagan, J., Lacey, B. C., & Moss, H. A. The visceral level: Situational determinants and behavioral correlates of autonomic response patterns. In P. H. Knapp (Ed.), *Expression of the emotions in man.* New York: International Universities Press, 1963.

Lewis, M. Infant attention: Response decrement as a measure of cognitive processes, or what's new, Baby Jane? Paper presented at the meeting of the Society for Research in Child Development, New York, March, 1967.

Lewis, M., Bartels, B., Campbell, H., & Goldberg, S. Individual differences in attention: The relation between infants' condition at birth and attention distribution within the first year. *American Journal of Diseases of Children,* 1967, vol. 113, 461-466.

Lewis, M., Bartels, B., Fadel, D., & Campbell, H. Infant attention: The effect of familiar and novel visual stimuli as a function of age. Paper presented at the meeting of the Eastern Psychological Association, New York, 1966. (a)

Lewis, M., Goldberg, S., & Rausch, M. Attention within the first year. Unpublished manuscript, 1966. (b)

Lewis, M., Kagan, J., & Kalafat, J. Patterns of fixation in the young infant. *Child Development,* 1966, vol. 37, 331-341. (c)

McCall, R. B. A stimulus-change theory of investigatory behavior: Statement and evidence. Unpublished doctoral dissertation, University of Illinois, 1965.

McCall, R. B. Discrepancy theories of exploration and affect: A proposed synthesis and evaluation. Unpublished manuscript, 1970.

McCall, R. B., & Garratt, C. R. Qualitative aspects of exploratory behavior in infants. Unpublished manuscript, 1969.

McCall, R. B., & Kagan, J. Stimulus-schema discrepancy and attention in the infant. *Journal of Experimental Child Psychology,* 1967, vol. 5, 381-390. (a)

McCall, R. B., & Kagan, J. Attention in the infant: Effects of complexity, contour, perimeter, and familiarity. *Child Development,* 1967, vol. 38, 939-952. (b)

McCall, R. B., & Kagan, J. Individual differences in the infant's distribution of attention to stimulus discrepancy. *Developmental Psychology,* 1970, vol. 2, 90-98.

McCall, R. B., & Melson, W. H. Attention in infants as a function of magnitude of discrepancy and habituation rate. *Psychonomic Science,* 1969, vol. 17, 317-319.

McCall, R. B. & Melson, W. H. Complexity, contour, and area as determinants of attention in infants. *Developmental Psychology,* 1970, vol. 3, 343-349. (a)

McCall, R. B., & Melson, W. H. Amount of short-term familiarization and the response to auditory discrepancies. *Child Development,* 1970, vol. 41, 861-869. (b)

McCall, R. B., Weiffenbach, J. M., & Tucker, L. A. Competing exploration and preadaptation in the course of light-contingent bar pressing over time. *Canadian Journal of Psychology,* 1967, vol. 21, 111-119.

McClelland, D. C., Atkinson, J. W., Clark, R. A., & Lowell, E. L. *The achievement motive.* New York: Appleton-Century-Crofts, 1953.

Melson, W. H., & McCall, R. B. Attentional responses of five-month girls to discrepant auditory stimuli. *Child Development,* 1970, vol. 41, 1159-1172.

Meyers, W. J., & Cantor, G. N. Infants' observing and heart period responses as related to novelty of visual stimuli. *Psychonomic Science,* 1966, vol. 5, 239-240.

Meyers, W. J., & Cantor, G. N. Observing and cardiac responses of human infants to visual stimuli. *Journal of Experimental Child Psychology,* 1967, vol. 5, 16-25.

Moffett, A. Stimulus complexity as a determinant of visual attention in infants. *Journal of Experimental Child Psychology,* 1969, vol. 8, 173-179.

Moore, T. Language and intelligence: A longitudinal study of the first eight years. *Human Development,* 1967, vol. 10(2), 88-106.

Munsinger, H., & Kessen, W. Uncertainty, structure, and preference. *Psychological Monographs,* 1964, vol. 78, (Whole No. 586).

Munsinger, H., & Kessen, W. Stimulus variability and cognitive change. *Psychological Review,* 1966, vol. 73, 164-178.

Munsinger, H., & Weir, W. Infants and young children's preference for complexity. *Journal of Experimental Child Psychology,* 1967, vol. 5, 69-73.

Ottinger, D. R., Blatchley, M. E., & Denenberg, V. H. Stimulation of human neonates and visual attentiveness. *American Psychological Association, Proceedings of the 76th Annual Convention,* 1968, Pp. 355-356.

Pancratz, C. & Cohen, L. Recovery of habituation in infants. *Journal of Experimental Child Psychology,* 1970, vol. 9, 208-216.

Piaget, J. *The origins of intelligence in children.* New York: International Universities Press, 1952.

Rubenstein, J. Maternal attentiveness and subsequent exploratory behavior in the infant. *Child Development,* 1967, vol. 38, 1089-1100.

Saayman, G., Ames, E. W., & Moffett, A. Response to novelty as an indicator of visual discrimination in the human infant. *Journal of Experimental Child Psychology,* 1964, vol. 1, 189-198.

Salapatek, P. Visual scanning of geometric figures by the human newborn. *Journal of Comparative and Physiological Psychology,* 1968, vol. 66, 247-258.

Salapatek, P., & Kessen, W. Visual scanning of triangles by the human newborn. *Journal of Experimental Child Psychology,* 1966, vol. 3, 155-167.

Schaffer, H. R., & Parry, M. H. Perceptual-motor behaviour in infancy as a function of age and stimulus familiarity. *British Journal of Psychology,* 1969, vol. 60, 1-9.

Sheldon A. B. Preference for familiar vs. novel stimuli as a function of the familiarity of the environment. *Journal of Comparative and Physiological Psychology,* 1969, vol. 67, 516-521.

Siqueland, E. R. The development of instrumental exploratory behavior during the first year of human life. Paper presented at the meeting of the Society for Research in Child Development, Santa Monica, California, March, 1969.

Sokolov, E. N. *Perception and the conditioned reflex.* S. W. Waydenfeld (Trans.). New York: Macmillan, 1963.

Spears, W. C. Assessment of visual preference and discrimination in the four-month-old infant. *Journal of Comparative and Physiological Psychology,* 1964, vol. 57, 381-386.

Thompson, R. F., & Spencer, W. A. Habituation: A model phenomenon for the study of neuronal substrates of behavior. *Psychological Review,* 1966, vol. 173, 16-43.

Uzgiris, I. C., & Hunt, J. McV. A longitudinal study of recognition learning. Paper presented at the meeting of the Society for Research in Child Development, Minneapolis, 1965.

Watson, J. S. Operant conditioning of visual fixation in infants under visual and auditory reinforcement. *Developmental Psychology,* 1969, vol. 1, 508-516.

Weizmann, F., Cohen, L. B., & Pratt, R. J. Novelty, familiarity, and the development of infant attention. *Developmental Psychology,* 1971, vol. 4, 149-154.

White, B. L. Child development research: An edifice without a foundation. *Merrill-Palmer Quarterly,* 1969, vol. 15, 49-80.

## DISCUSSION

If the same notes are used repeatedly as the auditory stimulus with only their order changed, how do you distinguish between fatigue and habituation in the response, a participant asked. Dr. McCall responded: If the infant displays an increased response to the presentation of the discrepancy, a minimal response to the next familiar standard, followed by a resurgence to the next discrepancy, etc., one can have a reasonable amount of confidence on the basis of this specificity of response that at least a part of the habituation phenomenon is linked directly with the stimulus per se and not simply and solely a function of general fatigue.

When we think about individuals being motivated by some sort of stimulus discrepancy, we might think about the individual and the stimulus being located along a common continuum, a participant suggested. We need to build into the technique some guarantee that the mediating dimension for the infant is some experimental continuum. Many investigators simply postulate the dimension. The alternative and perhaps more productive approach is to employ varying scale models. From a methodological standpoint, there are advantages in using multiple stimulus presentation procedures rather than the single stimulus presentation techniques Dr. McCall favors. It is possible to compare the relative magnitude of the preference ordering by presenting the stimuli sequentially to one group and in pairs to another group. The nature of the judgment process of the infant is apt to be different for two stimuli than it is for one.

To establish a scale criterion, you have to have some sort of transitivity in the order relation. For example, suppose we have three stimuli. If A is preferred over B, and B is preferred over C, the empirical question is whether A is preferred over C. Just getting an order relation on a series of stimuli does not answer the question of what the stimulus dimension is, but it moves us a little closer to the reality of the situation. If the experimenter can demonstrate that a particular dimension exists which determines the attention of the infant, there are some clear quantitative advantages; we can specify the magnitude of the discrepancy, which in turn helps to specify the shape of the preference curve. Furthermore, different models of discrepancy carry implications for the kinds of data one should get in this series of ordered relations. By building in sufficiently tight tests so that we can demonstrate that a particular continuum underlies responses, we will have a considerable amount of information which will help us specify discrepancy measures and the shape of the preference function.

In the 4 years he has been working on this research, Dr. McCall said, the most difficult task has been establishing a stimulus dimension; and he has not yet developed one that satisfies all the requirements he would like to place upon it. A simple ordinal scale is not entirely satisfactory, and no theory has yet laid out any specifications that will direct the construction of a scale possessing more than ordinal properties. Moreover, the dependent variable of cardiac change does not appear to be much more than ordinal. With respect to presentation methods, it is impossible to use cardiac change when stimuli are presented in a paired comparison situation because the infant may look back and forth between the two stimuli several times before the heart rate begins to change. Further, there is no way of knowing to which stimulus the cardiac change should be attributed, since its latency is so long relative to the speed with which an infant can shift his gaze.

Our basic purpose in this conference is to think about self-regulatory behavior. The selection of stimuli to orient oneself toward is obviously one of the most important features of self-regulation. So it is very important for us to know what determines his choices of what to look at.

Some studies have shown a switch at about 3 months of age from preference for the familiar to preference for the strange, another participant said. It seems possible that one explanation of this may have to do with the degree of discrepancy the child perceives; things may look less strange to an older child than to a younger one.

Is there any evidence about conditions under which the child will deliberately choose the familiar? In the Harlow work with monkeys, the animals would open a window to look through and see their familiar mother surrogate. But they would also open it to look at a brand new landscape. Kagan found that children would look both at scrambled faces and at normal faces; but they would smile more often at the normal, familiar faces. What conclusions can we draw from existing data? Are there instances in which the discrepancy hypothesis by itself is not sufficient to explain the behavior? What else besides discrepancy might make a stimulus attractive? Would high familiarity sometimes serve this function, if the object had been a source of needed resources?

Of course, Dr. McCall replied. Differential attention can be based on familiarity. For example, faces may bear a resemblance to mother, who presumably supplies nurturance and is associated with other positive circumstances. In his own work, however, he has elected to use sets of stimuli which do not have *a priori* power to recruit different amounts of attention. What they seem to show is that magnitude of discrepancy as he has defined it, is a variable. It is possible as has been suggested, that very young infants have a tighter discrepancy curve than older infants; that is, they avoid extreme discrepancies more than do older children. Such a proposition may assist us in understanding the interactive effects of age and familiarity. In terms of other variables, some studies of the fear of strangers indicate that there are certain stimuli that evoke a fear response even without prior familiarization, such as a looming object.

Further, in some circumstances, it may not be the stimulus itself that provokes fear but the uncertainty of needing to make a response but not knowing just which one.

Sheldon did some studies on object preference in rats, a participant said. When they were put in cages with a familiar and an unfamiliar object, they went to the familiar object during the early days of the trials but later began to go to the novel object. However, when the animals were placed in an unfamiliar environment for an hour before the final day's test, they tended to return to the familiar object when put back into the test cage. Could similar experiments be done to determine circumstances under which children would show preferences for familiar vs. novel objects? Would this perhaps show the child's preference is not just based on discrepancy but involves reward or reassurance of some kind?

A similar study by Weismann, Cohen, and Pratt has been done with human infants, Dr. McCall replied. The infant was experienced with a stimulus and tested for novelty–familiarity preference while placed in either a familiar or an unfamiliar bassinet. Unfortunately, the results showed a sex difference–boys attended more to the novel stimulus when in the familiar bassinet while just the reverse was true for the girls.

Is the basis of choice perhaps reinforcement, another participant suggested. At older levels, with children 4 or 5 years old, we see a great interest in quantitative concepts. At the supper table they ask, "Who has more?" When they listen to fairy tales, they want to hear over and over again stories like "The Three Little Pigs" which contain many quantitative gradations. Probably this is because such stories reinforce the quantitative concepts they are just mastering. Is it possible that some similar schema formation is involved in the preferences of very young infants? Dr. McCall agreed that infants and children do seem to spend much time "exercising" sensorimotor and cognitive behavior patterns, but in the cases mentioned it is difficult to separate the concepts of "functional autonomy," familiarity–novelty, "schema" formations, etc.

Perhaps some of the issues at stake would be clearer if a semantic obstacle were removed, another participant suggested. What Dr. McCall is studying, essentially, is looking behavior and heart rate deceleration. It confuses the issue to call this "attention." Looking behavior relates to attention, but it is not synonymous with it. Stechler has reported an experiment in which newborns were presented with a visual stimulus. While they were looking at this, another stimulus–an air stream on the skin–was presented. If these babies were really attending to the first stimulus, it seemed reasonable to assume they would show less response to the second stimulus. But this did not prove to be true; they showed greater response to the second stimulus. In a situation like this, we would not say the individual was not attending, but rather that he showed differential sensitivity. Perhaps it would be more accurate in relation to looking behavior to refer to "first fixation" rather than "attention."

This a point well taken, Dr. McCall agreed. It would be better at this point in our knowledge to talk about particular responses, but we need not avoid the proposition that, in some way yet unspecified, these behaviors may reflect aspects of the broader concept of attention.

# THE SENSORY-MOTOR ORIGINS OF KNOWLEDGE[1,2]

*Bärbel Inhelder*
University of Geneva

For some years now, we have been witnessing a prodigious growth of research into the beginnings of mental life, and scientific effort has tended to become increasingly centered on furthering our understanding of the developmental mechanisms themselves.

While the developmental psychologists had already progressed beyond the troublesome dichotomy of maturation versus learning, the advent of new theories of language, stemming from the linguistic work of Harris and, above all, Chomsky, have reopened the controversy between empiricism and nativism. Their studies have shown decisively that grammars, i.e., linguistic competence, have features which cannot be explained by recourse to the empiricist and associationist theories. Chomsky's work led him to make two hypotheses on the acquisition of language: first, that the basic linguistic structures are innate, and second, that when the child is in contact with the language spoken around him he makes a choice between the possible grammars (there are important restrictions on these grammars). His followers went further and stressed this rationalist aspect of linguistic theory; they saw in language the most important

[1] Translated by Mrs. S. Wedgwood (aided by a Ford Foundation grant).

[2] This paper was prepared for this meeting, but Dr. Inhelder became ill and was unable to present it so no discussion is included.

example of human cognition and from this evolved a neonativist argument going well beyond language itself.

As regards linguistic competence, this led to the hypothesis of a language acquisition device with a strong inner structure—a concept with which we can hardly quarrel. However, we begin to disagree when this structure is considered genetically preprogrammed and when no attempt is made to link the appearance of the first comprehensible utterances to the sensory-motor forms of behavior which precede their appearance.

Considering the question from the Genevan developmental angle, we think it is possible to trace continuous links between the first sensory-motor coordinations and truly cognitive structures, and to hypothesize that language and, in a more general way, the semiotic function of which it is the most obvious expression are not suddenly constructed but prepared by the elementary development of knowledge at the sensory-motor level.

Such a developmental point of view fits in with an explanatory system according to which the hereditary (both genetic and epigenetic) connections and the learning processes are controlled by more fundamental and more general mechanisms (e.g., regulatory and self-regulatory).

In the first part of this paper, we shall try to trace the development of the fundamental categories of knowledge (space, time, and causality) which, according to Piaget, are developed as early as during the preverbal period and form the basis of the fundamental operations and concepts of thought, culminating in the structures of scientific thought. We are trying to show that these first cognitive elaborations can be explained in biological terms such as assimilation; the development of intelligence is understood as a particular case of biological adaptation.

Second, we shall describe some current research which we hope will throw light on the passage from sensory-motor activity to the semiotic function such as it appears in the first forms of symbolic play and language.

We shall conclude with a few remarks on the heuristic value of a constructivist genetic concept which, of necessity, leads us to have recourse to self-regulatory models such as those found in modern biology and cybernetics.

## THE DEVELOPMENT OF THE UNIVERSE: THE ORIGIN OF THE CATEGORIES OF KNOWLEDGE

Certain remarkable studies of animal psychology, some of which have been transposed to the human level, have long convinced us of the existence of a sensory-motor intelligence which phylogenetically and ontogenetically precedes language. Nowadays, according to the epistemological perspective, the Kantian problem of the origin of knowledge has been placed in a context of developmental psychology. From this point of view, it is very interesting to follow step by step the child's progressive construction of his practical knowledge of the universe, a construction which has been compared to a miniature Copernican revolution.

In fact, knowledge begins in a sort of adualism, as J. M. Baldwin pointed out long ago. It does not stem uniquely from a subject already conscious of himself nor from an object which the subject simply accepts without question. It results from interaction between the subject and the object, and the progress of knowledge depends on the construction of modes of interaction or mediation.

Baldwin also showed us that the neonate manifests no awareness of his own person, nor of a stable boundary between the internal and external worlds until the construction of the ego becomes possible both in correspondence and in opposition with that of other people, at which time we now know, thanks to Piaget, that the child's primitive world begins to contain the first permanent objects. The first objects to be endowed with permanency are in fact the people closest to the child who, as objects, are particularly interesting.

At the outset, the neonate considers everything in relation to his own body as if he were the center of the world, but he remains totally unaware of this fact. During the first 18 months the revolution consists in decentering his self-centered actions, and culminates in the child's considering his body an object among others in space.

He gradually becomes able to link his actions and discover that he is the source or even master of his movements.

The progress which characterizes the constitution of the universe during the sensory-motor period is due to the progressive coordination of the subject's actions and their relationship with the displacements of objects in the immediate environment. The undifferentiation of the self and the outside world and the centration on his own body stem from the noncoordination of the actions: each action at the origin constitutes a small but isolatable whole and these actions directly link the body to the external objects (sucking, grasping) whence the undifferentiation between the subject and the objects. Later on, the coordination of his actions will dissolve this direct link and objects can acquire permanency; the subject then becomes capable of coordinating their movements into a coherent system.

Piaget's longitudinal observations on his three children, since confirmed by some excellent studies of large groups of children carried out by Gouin-Decarie, Escalona, Hunt, and Lézine, enable us to reconstitute the stages of this construction of the universe. We shall limit ourselves to describing the interdependency of the various aspects of this development.

During the period from about 6 to 8 months, in which the maturation of the central nervous system is sufficiently advanced to allow the indispensable coordinations of vision, prehension, hearing, and smell, we witness the constitution of the first systems of spatio-temporal and causal relationships.

These reveal a certain understanding of the universe, which, as it results from the subject's actions, is still subjective.

The beginning of object permanency is clearly linked to the child's activity:

Jacqueline at 0;8 takes possession of my watch which I offer her while holding the chain in my hand. She examines the watch with great interest, feels it, turns it over, says "apff", etc. I pull the chain; she feels a resistance and holds it back with force, but ends by letting it go. As she is lying down she does not try to look but holds out her arm, catches the watch again and brings it before her eyes. By contrast, if,in front of her I hide the watch without her having an opportunity to handle it, Jacqueline does not react and seems to forget it immediately.

Laurent at 7 months loses a cigarette box which he has just seized and thrown away. He touches it involuntarily outside his field of vision. He then immediately brings back his hand to where he can see it and looks at it for a long time, with an expression of surprise and disappointment—something like a feeling of loss. But far from considering this loss irreparable, he starts to swing his hand, even though empty, after which he looks at it again . . . as if he hopes that the object has been brought back.

At this level, the child is capable of following the movement of objects and even of anticipating rapid trajectories, but everything happens as if for him objects were not provided with autonomous movements.

Laurent at 7 months is sitting in his cot opposite my office door. I open the door, appear and make him laugh, then I move slowly to the back of the room; Laurent follows me with his eyes, but even before seeing me disappear from his field of vision he turns back toward the door, and does this several times in succession.

When the moving object really disappears, the child does not credit it with the power to continue its course.

It is just the same for the reconstitution of temporal sequences. The child already clearly perceives the "before and after" of events: for example, the mother's footsteps and her movements when she prepares herself for nursing are immediately interpreted as an indication of forthcoming satisfaction, but the child cannot, as yet, reconstitute the history of phenomena outside his own activity.

Laurent at 8 months sees his mother come into the room and follows her with his eyes until she is seated behind him: he returns to his play activities, but turns round several times to see her again, despite the fact that no sound evokes her presence. However, when his mother gets up and leaves the room, Laurent follows her with his eyes until she reaches the door, then, immediately after her disappearance, again looks for her behind him in the place where she was before.

The spatio-temporal organization centered on the activity of the subject himself naturally leads to a causality concept of the magico-phenomenist type, according to which action can be efficient from a distance without spatio-temporal contact.

Jacqueline at 8 months is in her cot when the canopy is shaken without the child being able to see the cause of the movement. At first she seems a little afraid, but then shows her joy by sitting up and then falling back into her cot. When the phenomenon is repeated she follows it with acute interest. When the movement of the canopy stops she sits up again, staring at the canopy as if waiting for it to move.

Similar manifestations are observed when her father meows, or puts out his tongue, which amuses her greatly.

To make interesting spectacles recur, she uses a number of ways and means, such as sitting up and shaking her hand; it is as if she believed in the efficiency of her gestures and movements.

Toward the end of the sensory-motor period, i.e., around the middle of the second year, the different action systems, instead of constituting little self-contained systems, become coordinated, and a connection between methods and goals which characterizes intelligent behavior is attained. The child thus discovers the possibility of using instruments to reach objects that are too far away. He becomes aware of himself as the source of actions and knowledge, since the coordination of the various actions supposes an initiative which goes beyond the immediate fusion between his own body and the external world. To coordinate actions to attain a specific goal or result supposes the progressive understanding of how to move objects and of how they can move autonomously. The child now behaves as if these movements comport systems, which are in a certain sense, as Piaget has shown, isomorphic with Poincaré's group-like geometrical transpositions. The child becomes capable of assigning a number of successive, determined positions to objects. The latter acquire a certain spatio-temporal permanency, which in turn "de-subjectifies" the causal relationships. In a way, this differentiation between subject and object on the level of activity results in the objects' acquiring a status of their own. This decentration places the child's own body, as an object among others, in a universe with a certain temporal and causal consistency.

In fact, Piaget's observation of his three children's behavior, at a slightly later age, reveals a practical understanding of the permanency of objects whatever their movements. The children manage to reconstitute the trajectories even when the objects themselves become and remain invisible, as shown by Jacqueline's search for an object:

At 1;7, Jacqueline is seated in front of three objects A, B and C, underneath which something can be hidden; these objects, in this case a beret, a handkerchief and her jacket, are equidistant from each other. I hide a small pencil in my hand saying: "Cuckoo the pencil," I show her my closed hand which I put under A; I then show it to her again before putting it under B; I then show it to her a third time before putting it under C where I leave the pencil and afterwards show her my open hand, repeating "cuckoo the pencil." Jacqueline immediately looks for the pencil under C. She finds it and laughs. When I change the object, its hiding places and their order, e.g., CAB, etc., Jacqueline succeeds nine times in finding the object.

It is clear that the child does not only remember the last position; proof of this is given by the fact that he sometimes tends to go back to the place where he saw the object disappear for the first time and that he temporarily relapses into less developed behaviors when the task becomes complicated.

Another example will convince us that the child is now capable not only of reconstituting the movements but also of anticipating them.

Jacqueline at 1;8 arrives at a closed door with a blade of grass in each hand. She stretches out her right hand towards the knob, but sees that she cannot turn it without letting go of the grass. She puts the grass on the floor, opens the door, picks up the grass again and enters. But when she wants to leave the room, things become complicated. She puts the grass on the floor and grasps the doorknob. But then she perceives that in pulling the door toward her, she will simultaneously push away the blade of grass which she placed on the floor. She therefore picks it up in order to put it outside the door's zone of movement.

As regards the reconstitution of small temporal series, we have several examples where the child reproduces a series of events and starts to evoke them by a beginning of symbolic representation.

At 1;7, Jacqueline picks up a blade of grass which she puts in a bucket as if it were one of the grasshoppers which a small cousin had brought her a few days before. She then says "Totelle, totelle, hop-la, boy" ("totelle" meaning "sauterelle," the French word for "grasshopper"). The perception of an object which reminded her symbolically of a grasshopper enables her to evoke past events and to reconstitute their order.

Finally as regards causality, this period is rich in important discoveries.

Laurent, at 1;4, tries to open a solid garden gate but does not manage to do so because the gate is held back by a piece of furniture. He cannot see why the gate will not open nor does any sound give him a clue, but having tried to force it, he suddenly seems to understand: he goes round the wall to the other side of the gate, moves the chair that was in the way and opens the gate with an exclamation of triumph.

We thus see that the coordination of the subject's actions is inseparable from the spatio-temporal and causal coordinations he attributes to reality; these two sorts of coordination together form the source of differentiation between the pole of the subject and that of the object. This decentration tending towards bipolarity paves the way, on the level of the action, for rational thought, and forms the basis of the knowledge categories of space, time, and causality.

How do we explain, in psychological terms, this progressive construction which we witness during the first 2 years of life?

Those who support learning theories based on behaviorist methodology tend to invoke connections of the associationist type, while Piaget explains these same phenomena biologically in terms of assimilation. In fact, the concept of association refers essentially to the external links between elements associated in time and/or space, while the concept of assimilation implies the integration of new data in an already existing structure or even in the construction of a new structure in the forms of schemes.

As regards the very elementary and uncoordinated actions, two interpretations are possible. In the first, the structure is already in existence, being hereditary: for example, the sucking reflexes. In this case, the assimilation consists only in the incorporation of new objects in the structure which belongs to the organic programmation. In the second, the activity is not preprogrammed: for example, the baby tries in vain to grasp a suspended object, then he succeeds in just touching it and it swings, which he finds a most interesting and unexpected spectacle. When the spectacle stops, he tries to reproduce it by what Piaget calls a reproductive assimilation (to repeat the same gesture) which results in the formation of a scheme. In the presence of another suspended object he will assimilate it to this same scheme; this implies an assimilation by recognition and, when he repeats the action in this new situation, a generalizing assimilation.

The coordination of actions by reciprocal assimilation represents both a novelty in relation to what preceded it and an extension of the same mechanism.

In the progress towards the elaboration of new behaviors we can thus distinguish two successive stages. A first stage is above all an extension of what is already known: it consists in assimilating simultaneously one object to two schemes, which is a beginning of reciprocal assimilation. If, for example, a swinging object also makes a sound, it can become in turn (or simultaneously) something to look at or listen to; a reciprocal assimilation can then lead, for example, to shaking any toy to find out what noise it makes. At first, goal and means remain relatively undifferentiated, but at a more advanced stage, the baby assigns a goal to his act before carrying it out: he uses various assimilatory schemes in order to produce sounds; for example, he pulls the canopy cord.

Such assimilatory processes will progressively result in the construction of new combinations. These discoveries stem at the same time from knowledge of the objects themselves (a suspended object is something which can start swinging) and knowledge resulting from action-schemes which the subject uses on them (coordinating the means and goals while respecting the order of succession of the movements to be carried out).

We thus witness as early as the sensory-motor level that the growing differentiation of the subject and the object is marked by the formation of action coordinations, but among the latter we can, from the beginning of mental life, distinguish two types; on the one hand, those which link the subject's actions, one to the other, and on the other hand, those which concern the effect of actions on the objects and vice versa.

The first, thus, consists in combining or separating certain action-schemes, in embedding them one in another, or ordering or putting them in correspondence, etc. They constitute the first forms of these general coordinations which are at the root of the logico-arithmetic structures whose further development at the level of thought will be considerable. The second confer a spatio-temporal, kinematic organization to the objects and this organization is the point of departure of the causal sequences.

Even before the appearance of language, therefore, we see that the coordinations of the sensory-motor actions result in the structuring of knowledge with its bipolar (logico-mathematic and physical) nature. The subject and the objects thus start to become differentiated when the modes of interaction between subject and object become more refined. But at this level the modes of interaction are still only concrete actions carried out by the child, and it is only thanks to the semiotic function that these actions progressively become thought-operations.

## FROM SENSORY-MOTOR ACTIVITIES TO THE SEMIOTIC FUNCTION

We still do not know enough about the links between sensory-motor and symbolic (or semiotic) behavior. Symbolic behavior is sometimes considered to have different origins from sensory-motor behavior—the two slowly converging; other psychologists consider that a continual process of interiorization takes place. While we favor this second working hypothesis, it is a rather weak one since it tells us no more than that processes similar to sensory-motor development take place without any external manifestations.

The common sense answer maintained that it was through the use of language as a system of socialized signifiers that the cognitive functions become free from the sensory-motor context and accede to the level of representation. However, since Pierce introduced to psychology the concept of the symbolic function and Head showed its importance in neuropathological studies, we can no longer consider language an isolated phenomenon, but rather a particularly important aspect of a more general function. Piaget's work, although not specifically concerning the beginning of language, has revealed the existence of a developmental interdependence between the various forms of the semiotic function, i.e., deferred imitation leading to imagined representation, symbolic play, and the first verbal schemes. In addition, he was the first to give a psychological interpretation of holophrases. All semiotic behaviors have in common the differentiation between signifiers and what they signify. While at the sensory-motor level signals cannot be separated from the actual objects or events, at the level of the semiotic function the signifiers are distinct from the context and the differentiation is introduced by the subject himself. A child who makes a shell slide along a box, saying "meow" knows very well that his shell is not a cat, but symbolizes one and that the box is not a wall, but represents one. Even if at first he regards the word "cat" as an intrinsic part of the animal, the word as such is for him a symbolizer (or signifier) and not the object itself. It is not surprising that the different aspects of the semiotic function first appear almost simultaneously in the same child.

### *Deferred Imitation*

Jacqueline 1;4 has a visit from a little boy of 1;6, whom she sees from time to time, and who in the course of the afternoon gets into a terrible temper. He screams as he tries to get out of his play-pen and pushes it backwards, stamping his feet. Jacqueline stands watching him in amazement, never having witnessed such a scene before. The next day, she herself screams in her play-pen and tries to move it, stamping her foot several times in succession.

The imitation of the whole scene is striking. Had it been immediate, it would naturally not have involved representation, but coming as it did after an interval of more than twelve hours, it must have involved some representative and pre-representative element.

### *Motor Symbolizer*

At 1;4, Lucienne tries to grasp a watch chain which she has seen being put into a match box which she does not know how to open. The opening is reduced to 3 mm. As a result of her previous experiences she possesses only two schemes: turning the box over in order to empty it of its contents and sliding her finger into the slit to make the chain come out. She immediately tries these two processes which fail. A pause follows during which Lucienne manifests a very curious reaction, bearing witness not only to the fact that she tries to think out the situation and to represent to herself through mental combination the operations to be performed, but also to the role played by imitation in the genesis of representations: She mimics the widening of the slit. She looks very carefully at it, then several times in succession, she opens and shuts her mouth, at first slightly, then wider and wider. She wants to enlarge the slit. The attempt at representation which she thus furnishes is expressed plastically, that is to say, due to inability to think out the situation in words or clear visual images she uses a simple motor representation as signifier or symbol.

Immediately after this phase of plastic reflection, Lucienne unhesitatingly puts her finger in the slit, pulls so as to enlarge the opening and grasps the chain.

### *Play*

At 1;3, Jacqueline sees a cloth whose fringed edges vaguely recall those of her pillow; she seizes it, holds a fold of it in her right hand, sucks the thumb of the same hand and lies down on her side, roaring with laughter. She keeps her eyes open, but blinks from time to time as if alluding to closed eyes. Finally, laughing more and more, she cries out "Nono" ("Nono" meaning sleep). The same cloth starts off the same play behavior on the following days. Later it is the tail of her donkey which evokes the fringes of the pillow and from the age of 1;5 onwards she makes her animals, a bear and a dog, also do "nono."

### *Language*

At 1;5, Laurent says "aplu" ("n'a plus" meaning literally "no more" or "all gone") signifying first of all a departure, then the act of throwing an object onto the ground; he then applies it to an object which is knocked over but does not disappear. He thus says "aplu" to his blocks when he knocks them over. Later "aplu" simply indicates that something is moved away (outside his grasp), then the game of giving someone an object to throw back to him. At 1;6, he even says "aplu" when another person has an object in his hand which he wants. Finally at 1;7 "aplu" becomes the synonym of recommencing any interesting play or action.

The development of the signifiers has been described by Piaget in functional terms. The development of the intelligence at each of its principal stages culminates in states of equilibrium between the two opposing and complementary tendencies of assimilation and accommodation. In the formation of the signifiers, sometimes one and then the other predominates. The accommodating aspects are clearly predominant in imitation and images, while those of assimilation play the major role in symbolic play. In language, the two tendencies are present in proportions varying with the stage reached in its development.

However, one problem remains completely open. How, in general, does the child acquire the capacity of substituting signifiers for objects or events? This is important since it is these substitutions which make possible the first symbolizations, which in turn are one of the conditions of the acquisition of language.

One of the best ways of studying the origins of symbolization would appear to be the systematic observation of a large group of children between 1 and 3 years old, some of them being followed longitudinally. Our Paris colleagues, Irène Lézine and Mira Stambak, specialists in the study of very young children, have been good enough to associate my psycholinguist colleague, Hermina Sinclair, and myself in such a study. A group of 66 children between 12 and 36 months is currently being observed in nurseries, with a longitudinal examination (regularly once or twice a month) being carried out on some of them.

The technique referred to in this paper consists of an observation effected from behind a one-way screen and the material used is made up of objects and toys with which the child is familiar, such as a sponge, spoon, plate, mug, pot, broom, feather-duster, miniature baby's bottle, cloth, paper, books, mirror, doll, baby, and teddy-bear. The child is put in front of these objects on the floor and one of the experimenters draws his attention to each of them and encourages him to touch them and play with them.

During the period between 12 and 26 months, we witness a development in behavior in which we can distinguish three levels of growing complexity.

In our first, inevitably superficial, analysis of this corpus of minute observations, we chose three different aspects of behavior, which allow us to distinguish different types of activity at each level of complexity.

The first type of activity is orientated toward knowledge of the object—the child is discovering the properties of the objects he is handling.

In the second type of activity the child himself introduces an organization into reality, rather than discovering something about a particular object.

The third type includes all the "make-believe" activities. It is, however, only at the most advanced level that one can clearly distinguish these types; at the lower levels, they are still too closely intertwined. For this reason, we shall begin with a description of the most striking characteristics of the most advanced level.

At this level, we witness the appearance of a host of new activities which can be classified according to the three different types we have distinguished.

*Activities of the first type.* We observe that the children are beginning to handle the objects in the accepted, usual way, both as regards the choice of the object-agent and that of the object-patient.

For example, Peter, at 26 months, sweeps the floor with the broom, brushes his hair with the hairbrush, and dusts the book with a cloth.

*Activities of the second type.* The children begin to put objects together in groups which can be interpreted as based on a first "category" criterion.

At the same session, Peter carefully puts the broom and the feather-duster beside one another.

*Activities of the third type.* Several subcategories may be distinguished in this group of symbolic patterns.

(*a*) The doll and the teddy-bear are treated as partners in a game. Peter holds the mirror up to the doll's face, inclining it so that she can look at herself.

(*b*) The activity includes symbolic evocation of a missing object. Peter pours some water from the bottle (in reality empty) down his neck which he wipes with his other hand.

(*c*) The activity includes the symbolic substitution of one object by another. Having put the doll in the nursing position, Peter puts the broomhandle in its mouth as if he were giving it the bottle.

One of the general characteristics of this level is the organization by the child of long sequences of actions which are linked, although sometimes very loosely, by a common situation.

The same Peter takes the brush and, using his right hand, slowly brushes his fringe, then he starts brushing the doll's hair. Then he holds the mirror up to the doll's face so that she can look at herself, then again brushes first the doll's and then finally his own hair.

To discover which behaviors precede and pave the way for these more developed ones, we must examine the intermediary level, which in our population is found at between 15 and 19 months of age. The following are some of the activities (still classified according to the same three types) which were witnessed.

1. The children carry out a group of activities (this time with less obvious connections between them) in which one of the objects, the agent, already plays its normal role while the patient-object is still more or less devoid of specific significance.

Peter, now aged 19 months, dusts the baby's face with the feather-duster. He puts the spoon on the plate and then presses it into the sponge. He brushes the baby's bottle with the hairbrush. Thus the action is performed with its usual instruments, but applied to an unusual object.

2. The children put various objects together, but at this level it is more difficult to distinguish a "category" criterion; these groupings have spatial, "category," and functional characteristics which cannot be isolated.

Peter puts the spoon in the mug and, using his left hand, presses it well in, takes the mirror with his right hand, knocks the spoon into the plate, puts the mirror into the mug then takes it out, puts the spoon back into the mug, then puts the spoon on the plate, puts the mirror into the mug, puts the spoon back into the mug, and finally puts the spoon back onto the plate.

Sometimes the spatio-temporal and causal aspects of the behaviors are more accentuated.

Peter takes the spoon with his right hand and puts it on the floor, stands it up vertically, swivels it round flat on the floor, stands it up again with his right hand, makes it turn around in this position, puts it on his right leg, then under his leg and then taps the floor with it.

*Activities of the third type.* At this level we can distinguish the following sub-categories:

*(a)* Symbolic activities using the child's own body.

Peter, again at the age of 19 months, rocks himself and pretends to sleep holding his head with his hand, his eyes closed. He stays like this for 2 minutes, getting more and more in a hunched-up position. As soon as the observer approaches, he opens his eyes and starts laughing.

At another time, he pretends to eat without food, using the spoon and the plate.

*(b)* Play activities making use of the doll or the teddy-bear, but without these being treated as active partners.

Peter hugs the bear and gently rubs the doll as if he were caressing it.

At this developmental level, the sequences are shorter and less well integrated into a coherent framework of organised activities.

At the most elementary level examined in this research, which, in our population is situated between 12 and 15 months of age, the activities are even more disjointed. It seems to be the objects themselves which trigger the motor schemes and almost any motor scheme can be applied to almost any object.

*Activities of the first type.* We observe an "instrumental" activity which prepares the way for the discovery of the properties of the object.

Peter, at 12 and 13 months, carries out a number of activities using two objects, such as hitting the mug with the broomhandle and pushing the pot with the feather-duster.

The activities using only one object are: shaking the mug, shaking the feather-duster (without our being able to recognize a precise imitative activity), throwing, rubbing, pulling and pushing one of the objects at hand.

*Activities of the second type.* The child puts one object on top of another (for example, the bottle on a plate, the mug on the mirror), into another (for example, the mirror into the pot), and beside another (for example, the spoon beside the broom).

*Activities of the third type.* No such activities can be observed at this first level. It is at this age that we see the most activities using the child's own body, but without being able to recognize a symbolic scheme, in the true sense of the term.

For example, Peter puts the bottle on his leg, bites the broom, scratches himself, pulls his hair, and explores the objects by putting them into his mouth; at this age, this last way of learning about things is still important.

In what way does this rather rapid examination of behaviors throw new light on the constitution of the symbolic function? Once the sensory-motor cognitive adaptation is complete, what necessary conditions have to be fulfilled before the symbolic function can develop?

As regards the "object-knowledge" pole, it seems that the object may be evoked in its absence or replaced by a signifier only after the child has endowed it with properties giving it significance.

As regards the subject's structurating activity, the detachment of the schemes from the material context can take place only after his activities have been organized into sequences within a coherent framework.

There is always, of course, close interaction between these two poles of activity, since the subject's structurating activity makes possible the constitution of the object and since the differentiation of the former is a function of the characteristics of the objects and especially of their kinematic properties.

Our viewpoint, as regards the first verbal productions, is inspired by the Piagetian concept that the first utterances are the translation of schemes whose subject- and object-poles are indissociable. Piaget calls these schemes "action judgments." Using the terminology of George Miller and Gruber (who consider the distinction between topic and commentary and the possibility of a topic becoming a commentary and vice versa to be one of the fundamental characteristics of language) it seems to us that the first holophrases represent a fusion between topic and commentary. By contrast, the slightly more evolved utterances, which Piaget calls "observation judgments," already reveal a distinction between topic and commentary (which does not imply that the two are necessarily present). At this same period, the child begins to refer to familiar objects in more or less stable and conventional terms.

It is possibly this aspect of knowledge, which we have called "discovery of object-properties," that will result in the acquisition of an organized lexicon; syntax, on the other hand, has its roots in the type of knowledge gathered from the subject's own organized activity. Evidently, there is an interdependence between these two aspects and each is constituted as a function of the other.

The first results of this type of research seem to give some weight to the hypothesis of a developmental construction on an interactionist model. We should like to emphasize the fact that when the development of sensory-motor activities is seen in the light of regulatory, rather than associative, mechanisms, it is easier to show its continuity with subsequent symbolic activities. The two aspects of action, one tending towards the object-pole and the other towards the subject-pole, cannot be interpreted without reference to the hypothesis of regulatory mechanisms. The latter cannot be rendered explicit through naturalistic observation and experimental studies have yet to be carried out.

## CONCLUSIONS

### *Developmental Model*

A developmental theory which is neither empiricist nor neonativist (but, in a sense, intermediary between those two positions) in the interpretation of the progressive construction of the cognitive functions, sooner or later leads to the invocation of feedback-organizing mechanisms. It is only these which provide a possible explanation for the subject-object interaction and the progress resulting from this interaction.

Piaget's fundamentally biological conceptions have led him from the outset towards a model with regulatory and even self-regulatory characteristics: a young cybernetician (Cellerier) was moved to say that Piaget became a cybernetician even before the science was invented. The heuristic value of Piaget's hypothesis received confirmation when he compared psycho-developmental factors with those of Waddington's embryogenetics (Piaget, 1967) and, in a more general way, with those of some of the contemporary biologists, e.g., Dobzhansky and Wallace.

In fact, according to Piaget, the regulatory systems are found on all levels of the organism's functioning, right from the genome up to psychological behaviors; thus they appear to be among the most general characteristics of the organism. Self-regulations seem to constitute at the same time one of the most universal characteristics of life and the most general mechanism to be found in both organic and cognitive behaviors. This is so regardless of whether we are concerned with what (at the level of the genome) Dobzhansky's school calls genetic homeostasy, with the dynamic equilibrations of the embryogenesis named homeorhesis by Waddington, with the multiple regulations of the nervous system including the reflex feedbacks or, finally, with the regulations and equilibrations observable at all levels of cognitive behavior.

Each self-regulatory system supposes the existence of several types of feedback, which are caused by interaction of the subject and the environment and by the changes in the environment itself.

In cybernetics one can consider that a self-regulatory system can be self-constructing. A system is modified through the influence of a feedback constituted by the results of its own action by which it treats the input coming from outside.

We hope we have shown in this paper some of the reasons why we feel that the development of the cognitive functions during the first phases of development and, in particular, the continuity between sensory-motor and semiotic activities can best be explained through the self-regulatory mechanisms.

## REFERENCES

Inhelder, B., Lézine, I., Sinclair, H., Stambak, M. *Les origines de la fonction sémiotique,* in preparation.

Piaget, J. *Play, dreams, imitation in childhood.* New York, Norton, 1951 (French edition, 1946).

Piaget, J. *The origins of intelligence in children,* New York: International University Press, 1952 (French edition, 1936).

Piaget, J. *The construction of reality in the child,* New York, Basic Books, 1954 (French edition, 1937).

Piaget, J. *Biologie et Connaissance,* Paris, Gallimard, 1967 (English translation in preparation, Edinburgh University Press).

# OF LANGUAGE KNOWLEDGE, APES, AND BRAINS [1]

*Eric H. Lenneberg*
Cornell University

## LANGUAGE KNOWLEDGE: FORMAL CONSTANTS OF COGNITIVE ACHIEVEMENT

Man's language ability is due to a more general, deep-seated cognitive ability characteristic of the species. It is argued that man's ability for mathematical thinking is a product of the same species-specific form of cerebration as language. The basis for mathematical constructs seems to be contained in the basis for language constructs; apparently, for every mathematical notion there is a homologous one in the sphere of language, the former always being more restricted and well defined than the latter. Mathematical ability may therefore be regarded as a special case of the more general ability that also generates language, and this point is further emphasized by certain similarities in the formal structure of mathematics (arithmetic in particular) and language.

Taking advantage of the commonalities between language and arithmetic, it is possible to use the latter to illustrate important general characteristics of the former. The insights gained are relevant to biology at large, and comparative zoology and neurology in particular. The zoologist who wishes to compare animal communication with language must know what the nature of language

[1] This paper has resulted from research carried out under Grant No. 2279 from the Wenner-Gren Foundation for Anthropological Research/New York, N.Y. It is reprinted by permission of the *Journal of Psycholinguistic Research* where it was published in Volume 1, pp. 1-29 (1971).

is–how (or whether) one might analyze it into components. He must know what might constitute a primitive or simple language. It is shown that the irreducible elements of the two systems under study (language and arithmetic) are *processes* (i.e., processes of "relating," or simply *relations*) and that these processes combine into integrated systems. The systems have ontogenetic histories that might, perhaps, furnish a criterion for the notion of simplicity. We do not yet know what might be a homologous phylogenetic "cousin" of the basic human ability under consideration; however, we should expect it to be "homeomorphic" to the human system if it is derived from a common ancestral ability. Homeomorphic mapping is therefore the most reliable criterion so far for phylogenetic relatedness.

By characterizing language and arithmetic simultaneously (in order to get at their common biological foundation) it is also possible to sharpen up the questions that the student of language should put to the neurophysiologist. The quest for innovation or differences in brain *processes* and *functions* now appears to be the primary one, whereas a description of structural changes in the human brain would be of interest only insofar as this would elucidate how brain functions might have become modified by them.

## Introduction

Every aspect of an organism's behavior bears the indelible imprint of the biological operating principles of its own species. Whatever a cat does, it does in a feline fashion; whatever man does, he does in a human fashion. This may be taken as an axiom (or truism, for that matter, for how could it be otherwise?). An animal cannot change its constitution between behaviors; every movement, every sensation, every insight, every motive is mediated, coordinated, regulated, transformed, integrated, etc., by one and the same nervous system, the same skeleton, the same irritable tissues. Despite the infinite variability in individual acts or skills (many of them dependent upon environmental circumstances), there remain limits to the behavioral variations, and there are physiological constants that are common to literally everything a given animal does. I would like to propose, as a general guideline for research in animal behavior, that one can gain important insights into the biological nature of a species by attempting to discover the physiological constants–by searching for the common denominators underlying the different sorts of behavior in an animal's repertoire.

In this article, I am concerned with man's capacity for language; however, I should like to deal with it in the context of his more general cognitive propensities. All of man's intellectual activities must bear the hallmark of the mode of cognition imposed by the operating principles of the human brain. In our present state of knowledge, it is hardly possible to discover the neurophysiological constants of all intellectual activities. But it may be possible to demonstrate that the activities themselves have a common denominator. If we choose an appropriate data language to describe our observations (hopefully a formal language, a logic or an algebra), it should be possible to point to *formal constants* in any two or more intellectual activities; these formal constants, in turn, may lead the way toward formulating biological hypotheses and eventually even toward neurophysiological discovery.

I shall point to some commonalities in man's capacity for doing arithmetic and his capacity for acquiring a natural language. The formal constant underlying these activities is a peculiar *relational structure.* One could go on to show that man's perceptual activities also share in this formal constant, but space does not permit elaboration of this point here (see Lenneberg, 1970).

Once the formal constant has become apparent, I shall try to apply this insight to some common questions that arise in the study of language—for instance: what constitutes primitivity in language; are language skills a conglomerate of independent traits or habits; how can one decide whether an ape has language; and how does one begin to look for brain correlates of language?

## The Capacity for Arithmetic

Knowing a language is not the same as knowing arithmetic; nevertheless, the two have much in common. Arithmetic behaves like language in several important ways. All of the primitive notions that form the foundation of arithmetic are part and parcel of any natural language, and the operations of grade school arithmetic are derivatives of language operations. Arithmetic "sentences" are much more restricted in syntax and vocabulary than are the sentences of a natural language, but arithmetic sentences and language sentences have similar formal properties. It is not difficult to imagine what sorts of tests you would use to examine an organism's knowledge of arithmetic; we shall see that these tests also provide us with a good experimental paradigm for testing the presence of language knowledge.

First, we would test for the presence of arithmetic knowledge in a subject by giving him problems to solve, such as $6 + 4 = ?$ If the subject failed in tasks of this sort, we would try to test for his capacity to understand the basic concepts of arithmetic. To do this, it would be necessary first to analyze arithmetic itself, in order to discover these basic concepts. Let us say that there is a set of elements, called cardinal numbers, and at least one operation, called addition. As a starter, it is tempting to argue that the operation is a more abstract notion than is the concept of numbers, and that we should, therefore, treat the capacity for learning numbers separately from the capacity for learning the operation of addition. Furthermore, there is a temptation to explain numbers in a nominalistic vein: a number is "simply the name that we have learned to associate with something we can see"; and counting is "the consequence of another somewhat similar convention, the learning by rote of a sequence" (like the alphabet).

There are three important reasons for not accepting these nominalistic explanations.

1. Although it is possible to see at a glance couplets, triplets, and quadruplets of things, man begins to make mistakes for collections larger than four. The mistakes increase in magnitude and frequency as the collection becomes larger.

2. It is conceivable that a subject could name a trio or a quartet of one set of objects correctly but would fail to see that three marbles have something in common with the three legs of a stool; knowing the *word three* does not ensure understanding of the *concept of three.* Let us call concepts such as *three, one hundred,* etc., classes (following G. Frege or B. Russell). The notion *class* does not imply here the physical existence of a collection, but merely the existence of some principle or criterion by which one may decide whether a given instance is or is not assignable to membership in some group. If I can assert of something that it is *green,* I have used a criterion by which things may be assigned to the class "green things." If I can assert of a collection that it is *three,* I have used a criterion by which any conceivable collection could be evaluated as to *threeness* (the collections may range from fresh eggs to beatitudes). Of course, we have yet to define what the criteria are for the classes called cardinal numbers. But first, back to the reasons for rejecting a nominalistic definition of numbers.

3. If a person knows the correct sequence of number words from one to ten, it does not follow that he can also count correctly seven of his fingers. For instance, many children recite the words and may even turn down one finger at a time, but when asked how many fingers they have counted after having turned down number seven, will give a random answer. The knowledge of the automatic sequence is clearly not the same as knowledge of how numbers are ordered. To say that counting is a "simple mapping operation" from fingers to objects begs the question of the nature of the number concept. A child is capable of a one-to-one mapping operation without having the number concept; he can, for example, take down the right number of plates from the cupboard for a company of six by giving each plate the name of one of the company. Thus, mapping is a relation that is more primitive than numbers. In short, the concept of numbers cannot be explained nominalistically or as a cultural convention; the number eight is not simply a visual pattern arbitrarily associated with something we can perceive directly.

What, then, is the knowledge of particular cardinal numbers? There is a sizable literature on this problem that need not be reviewed here (see Benacerraf & Putnam, 1964). It is possible to define any cardinal number if we allow at least the following three concepts to be intuitively obvious: the concepts of *one,* of *and,* and of *is.*[2] If we don't have to explain what these concepts are or how the subject comes by them, then we can say that we give the arbitrary name *one* to that which is perceived at a glance as a single. If we also know what the words *and* and *is* mean, we can say *one and one* is what we call *two; one and one and one* is what we call *three,* etc. Obviously, this isn't much of an explanation. Most disturbing of all is the question, what gives us the wherewithal to count the number of times we have said "and one"? At least we must introduce another notion, namely that the knowledge of the quantity *one* may be applied to the operation of adding the words *and one* to *one;* thus, if we know intuitively what one is, we can also know intuitively what two is, namely a single "and-one operation." Now that we have the general concepts of one and of two, nothing can keep us from applying these numbers to themselves; we can count numbers with numbers, such as one or two singles or one or two couples. If we hear the number two, we immediately understand that it may stand for two couples or for two singles. Notice that we had to add a fourth concept to the first three, namely *iteration,* which makes it possible rapidly to extend our number system *ad infinitum.* This concept answers the question "how often?" or "how many times?"; by counting the number of times, we are introducing multiplication, an offshoot of addition.

We still have to face another major difficulty, which should be easy for us to appreciate in our capacity as psychologists. What are the primitive concepts *one, and, is*? What would we have to do, for instance, to demonstrate that an animal knows what is meant by *one*? Clearly, we would have to show that he knows the difference between one and not-one, that is, one and many, singles and multiples. Harlow's oddity paradigm is an approximation of such a procedure, although in that paradigm it has not yet been demonstrated that the animal has, in fact, learned the property that underlies the entire class of singles, on the one hand, and multiples, on the other. I believe that it should not be difficult to devise an appropriate experiment on this problem, and we need not discuss the details here. I must emphasize, however, that one-ness is clearly a *relation,* as is many-ness; the classes of ones and not-ones are defined by that relationship. The notions *and* and *is* also imply relations.[3] Again, it should not be difficult to devise experiments in which the subject's capacity for understanding either of

[2] I am concerned here with psychological primitives, not with strict axiomatization such as, for instance, Peano's five axioms. Peano introduces the word *successor* axiomatically but makes no mention of the concept *and.* However, the notion *successor* is defined for a natural system as "any arbitrary number *plus* one." I believe it is correct, therefore, to regard the concept *and* as a psychological primitive.

[3] A short demonstration that *one, and, is* are relations may be in place. In referring to a bunch of mice in front of me, I may say either: there are *six* mice, or there are *three* sets of

these relations is tested. Imagine an organism that performs five tricks upon five different signals; can it learn to differentiate between the command "trick A *and* B" and the command "trick A *or* B"? An experiment by which the general comprehension of *is* could be tested might follow this procedure: A subject is trained to perform either of two tricks, A *or* B. The signals for the performance of the tricks are $x$ and $y$, where the only requirement is that $x \neq y$, but either of these variables may take on any value. Thus, one day $x$ is a high tone and $y$ a low one; the next day $x$ is a light and $y$ a gong; every day the stimuli change, but throughout the day's session remain constant. If the subject comprehends the notion *is,* he will never make more than one initial mistake, at the beginning of the day's session, because all he will have to find out is what today's $x$ "is." The logic here is "if $m$ means A and $n$ means A, $m = n = x$."

---

twins, or there are *two* sets of triplets, or there is *one* litter. It is clear that the concept *one* only makes sense relative to some other concept. Thus, it is intuitively acceptable to say that *one* implies a relation. Take some of the more "obvious" relations, such as *father of* or *bigger than.* These, as any other relations, necessarily imply (1) two sets of classes, $S$ and $M$, and (2) a mapping, $\rightarrow$ , connecting one or more elements in $S$ to one or more elements in $M$. The act of relating defines some order between elements of the respective sets (in the instance to be considered, one element in each set; therefore we speak of ordered pairs). For example: Jerry is father of Ben; Roger is bigger than Miriam. Thus the binary relations considered here "produce" pairs or, in other words, form a new set or a class of ordered pairs. In set-theoretical language this set of ordered pairs is a subset of the cartesian product of the two sets $(S \times M)$ presupposed by the relation. We see that a relation implies *(a)* a mapping, *(b)* the existence of two sets, and *(c)* the formation of a new set or class.

Notice that the two sets implied by every relation may actually come about by splitting up a single set. If we define the relation $R$ as *is one of,* then $R$ always implies such a splitting-up operation. Let I be the set of integers, then 5 $R$I, O $R$I, etc. The same holds true if I assert of anything that it is one.

Let $S$ be defined as $\{1, 2\}$ and $M$ as $\{0, 1, 2, 3\}$. The mapping

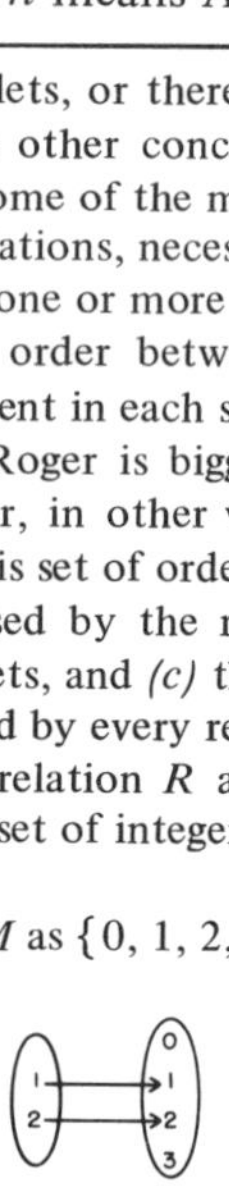


$S$ $M$

may be taken as the relation *is equal to.* Again, let $S$ be the set of ordered pairs $\{(1, 1)\ (1, 2)\}$ and $M$ be defined as $\{0, 1, 2, 3\}$, and consider the following three mappings:

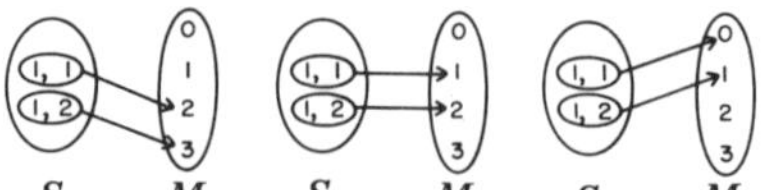


$S$ $M$ $S$ $M$ $S$ $M$

(a) (b) (c)

In what sense can these mappings be said to be true? They become true by specifying what the relation (or kind of order) is that holds between the elements of the pairs in $S$. Mapping (a) becomes true by making the relation be *and;* (b), by making it *times;* (c), by making it *difference between.* This shows that operations (such as addition, multiplication, subtraction, etc.) also imply relations.

We have made the following points:

1. The operation addition expresses a relationship.

2. Numbers stand for relationships, and cardinal numbers larger than one express relations between relations.

3. There is no definition in arithmetic that does not itself rely upon a relation, and thus arithmetic is a closed system with respect to relations.

4. The primitive relationships underlying arithmetic with cardinal numbers (1, +, =) can also be expressed in every natural language.

5. The presence or absence of a word denoting a relationship in the vocabulary of an individual is not correlated with that individual's capacity for understanding the given relationship.

It may be well to add here a parenthetical remark that will elucidate my argument in the following sections. Whenever we assert the existence of a relationship, we are implying that something was, is, or will be related to something else; this necessarily implies the existence of a correlated *activity* or *process.* Further, if we say of an organism or artifact that it relates something to something else, we are necessarily talking about some activity in the organism. If an organism can relate things in more than one way, if its output is more than a single relationship, it must be capable of more than a single sort of activity. One might say that such a "device" can be in different activity-states, depending on the particular relationship it is computing.

When we teach our children arithmetic, we introduce them at once to the set of natural numbers. If we were trying to teach arithmetic to some other form, say a chimpanzee, it would, perhaps, be easier to choose a system with a finite number of elements. A good example would be a residue class of, say, only four elements, 0, 1, 2, 3, such as shown in Figure 1. The table contains all the addition facts of the residue class (modulo 4). The logic of these additions is illustrated in the dial; it represents a number line turned back onto itself. The origin is zero, and the distance zero to one is unity. From the dial, we see immediately why three and one would be zero in this system, and three and two, one. The subject would have to be trained to point to any one of the four numbers in response to such questions as "what is one and two?" We might use ten of the addition facts for training and demonstration and use the remaining six facts to test his understanding of the underlying principles.

It is always difficult to give a "natural" definition of simplicity or primitivity. Perhaps addition with residue classes is a simpler or more primitive level of arithmetic than adding natural numbers because of the small number of elements. Similarly, the capacity for understanding any one of the relationships that underlie arithmetic (*one, and, is*) may perhaps be regarded as a "precursor" of arithmetic *capacity.* However, if an animal could operate with only one or two of these primitive relationships, one would hardly be tempted to say that it knew "a primitive form of arithmetic," since it could not even learn to count.

| + | 0 | 1 | 2 | 3 |
|---|---|---|---|---|
| 0 | 0 | 1 | 2 | 3 |
| 1 | 1 | 2 | 3 | 0 |
| 2 | 2 | 3 | 0 | 1 |
| 3 | 3 | 0 | 1 | 2 |

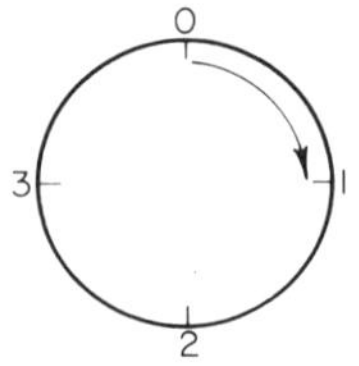


**Fig. 1.**

The Capacity for Language

Our aim is to show that language behaves like arithmetic in several essential ways. If we succeed, it will be easier to see that certain points that appear obvious enough when we are considering arithmetic knowledge apply equally well to language knowledge.

The study of algebra makes it most obvious that mathematics is concerned with pure relationships. For instance, we allow the symbols $a$, $b$ to stand for any numbers; then we relate them, such as $a + b$, $a = b$, $a/b$, etc. As long as we are doing pure mathematics, we are usually not concerned with either giving names to relational constructs or finding out the exact number of a particular relationship that we construct. In natural languages, we have the same situation. We can express pure relationships by letting the symbols *a, b* stand for any words, and then relate them by saying a *and* b, a *is* b, a *to* b, a *from* b, etc. Evidently the repertoire of relationships in language is more varied than it is in the realm of high school algebra.

In high school algebra, variables such as *a* and *b* are intended to represent numbers, where numbers stand for classes or sets. The situation in natural languages is quite analogous. Here the variables *a* and *b* also stand for classes, such as nouns, adjectives, verbs, etc. The elements in classes such as nouns, again, stand for classes—not for individuals.

We argued earlier against a nominalistic approach to the meaning of numbers. These arguments also apply to a nominalistic approach to the meaning of words. We can use the same three arguments advanced on page 160. If the variables *a* and *b* stand for words, their range is not defined by or limited to what can be perceived at a glance out in nature. Just as numbers represent quantities beyond those that can be perceived directly, words quite clearly can represent constructs that also cannot be observed directly (freedom, truth, brotherhood). Further, the word *shoe,* which seems concrete enough, does, in fact, stand for a class, a point that has been made so often before that it needs no elaboration. Shoe-hood is a construct, the result of some computation (that is, operations or relations; I don't care whether it is to be called stimulus generalization or a concept), and the word *shoe* stands for this computation or family of computations.

If someone knows the name of something specific, a proper name, he does not necessarily have the underlying capacity for understanding the principles upon which the classes are constructed to which this individual belongs. If the name of a dish with a certain pattern on it is "three" (or ding-ding-ding), it does not follow that the animal who has learned this relationship can construct the class of three-ness. Knowing that a certain man is called Pablo Casals is no indication that the knower can understand the class *cellist,* or the class *man,* for that matter.

A natural language can exist without any proper names whatever. And conversely, adding any number of proper names to a language does not change its essential structure; since *name* is an open set, there may be an infinity of names, for all we care. The naming of specifics seems to be quite irrelevant to knowing how a language works. This insight may even be extended to the naming or labeling of specific tricks, such as *heel, beg, show me your nose,* etc. If an infant or animal responds to a verbal command by one stereotyped act, the most important property of language is missing; because no relation is being tagged, there is no generality and no productivity. Formally, the situation is no different from a startle elicited by a bang (or an associated conditioned response), which could hardly be called a primitive form of a natural language.

At one time, it seemed to many of us as if a form of nominalism were introduced into language systems by attaching names to specific sensations; once sensations had been given names, all other words could be defined in terms of these names. There are, however, good reasons now to discredit this idea. First, we don't know what pure sensory qualities are (is *bigness* a sensory quality? is *loudness?*), and second, even in instances where this doubt is minimal, such as in the perception of colors, it may be shown empirically that the words relating to these sensations are always used as relative terms. The word *red* may apply to a narrowly circumscribed range of colors when we refer to a color chart, but when it comes to cows, we are not confused by calling one red, even if we see the difference between that color and the one on the color chart. Every sensory word fits into a system of related sensory words; it is the system that is learned and used, not individual and independent items from it. Color words illustrate this beautifully. As long as a child has just one or two words for color, he uses them rather randomly. But when he has four or five color words, his usage suddenly becomes normalized; the relational system has been established, and now the only difference between his usage and that of the adult is that his individual color categories are still somewhat wider than those in the adult system. A semantic study of the words for sensory experience makes it clear that a simple nominalism cannot explain even the use of sensory terms. There is no escape from the necessity of looking toward relations as the basis of meaning.

The third argument used in connection with numbers also applies to language: words by themselves neither define nor delimit nor are evidence for the presence of the underlying relational capacities that appear to be signaled by them. It is common that a child (or a blind person) uses color words without apparently knowing what the words mean. On the other hand, a sighted child may show a pronounced preference for red things but be unable to comprehend the meaning of the word red.

We found that in arithmetic, the classes labeled by cardinal numbers larger than one cannot be defined without introducing the elementary relation "and"; thus the construction of the elements in the system of cardinal numbers is intimately related to the addition operation defined for this system. Again, if we look at natural languages as a whole, we find a very similar situation. The elements in the set *natural language "L"* are words, and the operations are defined by various relationships that may be brought to bear upon the elements. Ideally, we would like to draw a sharp distinction between the lexicon, on the one hand, and syntax, on the other, such that syntax is concerned only with relationships, and the lexicon is simply an extensive class of items. However, no such dichotomy is possible. In a highly inflected language, practically all relationships are or at least can be expressed by the words themselves (or by some morpheme). And even in the so-called synthetic languages, such as English, some relationships are signaled by words and some by such syntactic devices as word order. If we look at historical changes in language, it is clear that no relationship whatever is in principle exempt from being taken out of the realm of pure syntax and being incorporated into the lexicon. As a matter of fact, the weight of my argument here has been that every word stands for relational constructs, and therefore the distinction between syntax, on the one hand, and lexicon, on the other, can never be one of relations or operations as against absolute items.

We discussed earlier the bootstrap operation by which the number system is constructed and expanded *ad infinitum.* It consists essentially of iteration of the addition operation and of letting the primitive concepts *one, and, is,* as well as the new concepts generated by means of them, apply to themselves, that is, letting numbers count the number of times the addition operation has to be repeated. Thus, *one* may be the number of a single count of ten, as in the decimal system. Language also is a system of open sets, and it can be expanded *ad infinitum* by iteration in ways formally similar to the ways in which the number system we have been considering is expanded. Words denote relations, and the relations between words can be denoted by words. Furthermore, any composition of words (a construct of constructs) can be replaced, in principle, by a word, much the way any composition of numbers can be replaced by a number. For instance, anything that has been said or written can be given a name such as *sentence* or *the complete Shakespeare* (abbreviated to *Shakespeare*), and thus be incorporated into any new composition as a single element. Mappings of this sort have much in common with the mappings implied by an arithmetic sentence. Although there are only certain instances in language in which a given relationship has a unique answer (the author of Waverley is Scott) as in certain arithmetic sentences (2 + 3 = 5), we find that any particular ordered or related pair may be equated to any of the elements of a particular set (like the solution set of a given mathematical equation); for instance, the relationship *young human being* stands in a similarity relationship to *infant, inexperience, issue of fertile parental union,* etc.

The most impressive property of simple arithmetic may be its application to everyday situations, like figuring out our change at the grocer's, and it is not clear that language has this particular property. There is no need here to ponder the academic question whether the usefulness of arithmetic is the same as or different from that of language (a somewhat foolish problem anyway). But it should be clear that the *nature* of arithmetic is not its application. Attempts to characterize arithmetic will not be based upon its applications, but upon a description of its peculiar relationships (and, of course, the relationships between them). Just as arithmetic is not "solving equations" or "figuring out percentages," but a system of relationships (the byproducts or benefits of which are the solutions of practical problems), so language is not essentially defined as communication or exchange of information. These are its byproducts, benefits, or applications, but the nature of language is a particular system of relationships, and language knowledge is the capacity to relate in specified ways.

In summary, then, we found (a) that both arithmetic and language are structures of relationships and that there is nothing in either system that can be defined without again making use of some relationship in the definition; (b) that at least the primitive relations of arithmetic (1, +, =) are also marked (either by a word or a syntactic device) in natural languages; (c) that neither of the two formal structures relies upon the use of symbols that stand for unique or specific events or objects; (d) that for both systems we encountered conceptual difficulties in giving mutually exclusive definitions of elements, on the one hand, and operations, on the other; (e) that an organism that "has words" does not necessarily have the understanding of the relational constructs that are customarily designated by these words.

Despite these similarities between arithmetic and language, there are, of course, some differences. Some of these are probably less important than others. Let me begin with one that may well be spurious. There seems to be an air of greater concreteness about language than about arithmetic. However, this may be due simply to man's peculiar mode of dealing with the world. For a baboon, language relations may be as abstract as arithmetic ones. And how are we to define *concrete* and *abstract* in objective terms? These notions are as relative as the notions *simple* and *complex.* The appropriate assignment of objects into classes such as *trees, pebbles, mushrooms,* and *none-of-these* evidently requires the "recognition" or computation of invariances that are as abstract as any notion in geometry or topology.

A more serious difference is that the number of relations in a natural language is clearly larger than in arithmetic, where it is possible to take an exhaustive inventory of all the relations admitted. The primitive relations of arithmetic are, at least potentially, available in language, and we may assume that they come from a universe of relations that is the universal set for both arithmetic and language. Although a natural language is limited by certain conventions, linguists believe (rightly so) that anything may be named in any natural language, and on this assumption, any natural language is potentially larger in terms of the relations it contains than any mathematical system. It is therefore reasonable to believe that arithmetic is a special structure within the more general structure of language.

Although we can say that *one, and, is* are basic, primitive relations in arithmetic, no one has yet succeeded in postulating the minimal set of primitive relations necessary for the construction of a natural language, nor even in making a reasonable list of "simple, elementary relations" (no matter whether "necessary" or not for a language). This gives natural languages a certain undefinability and amorphousness.

Closely related to this point is the lack of regimentation in natural languages. Whereas all relationships in arithmetic are clearly defined, particularly those between symbols and concepts, language indulges in frank sloppiness in this respect. Nothing is well defined, not even the relations between symbols and concepts. Even the most elementary relation words in natural languages usually have more than one use, and thus seem to represent a variety of relations. For example, the word *is* may indicate 1. predication-attribution, 2. similarity, 3. equality, 4. identity, 5. equinumerosity. (One may quibble with this particular list and substitute some other one for it; the point merely is the multiplicity of denoted relationships.) Similarly, the word *and* does not have the same denotation in the sentences: Peter *and* Paul are hungry; One *and* two are three. In the case of most nouns, verbs, and adjectives, the situation is even more pronounced. We might, for instance, give a list of relationships that constitute the denotation of the word *shoe* in a particular sentence and context. If we now were to compare this list with similar lists based on different occurrences and usages of the same word *shoe,* we would find that all the lists were somewhat different one from the other, although there would always be overlap. If we did the same thing in a historical perspective, we would frequently find that the lists have changed with time, so that often there is no overlap between old and new denotations. But even on a synchronous level, we may find that a word is spontaneously being used metaphorically, in which case there may be only a small number of relationships in common with the majority of lists. From this we obtain the following picture: there is a set of operations or relationships (determined by the capacities of the brain), which we can represent as a formal structure called a *space of relations.* Words can be mapped into this space, but points in the relation-space do not stand in a one-to-one or even a fixed relationship to words. Instead, words "hover" over the space and are capable of being mapped into a region of the space, which I shall call the semantic field of the word. However, the field is ill defined and has a way of sliding around in the space. For example, by changing word usage, the field shifts now in this, now in that direction, and, with the lapse of time (centuries), it may traverse considerable distances in the relation-space.

Because words map only onto ill-defined fields, the interpretation of words between speakers can never be exact. We know with only varying degrees of certainty what our interlocutor means by the words he is using. The efficiency of communication is not vitally affected by this sloppiness as long as the variance in denotation between the two speakers is not too great. In other words, we can *tolerate* a certain degree of shifting and fuzziness of the fields. This notion could be formalized, and the *tolerance upon the fields* could, at least in theory, be defined with some rigor (Zeeman & Buneman, 1968).

As in the case of arithmetic, the most general thing we can say about language is that it is a system of relations and that knowing language means relating something to something else in certain ways. Oddly enough, what is being related is still relations, a statement that might sound less odd if we could get ourselves to admit that knowledge itself must be understood as a process of relating.[4] Again, we are at a loss to say what constitutes simplicity or single elements in language knowledge. Undoubtedly, whatever it is, it must be certain types of relations, perhaps some of those that are found most widely among the languages of the world–predicate relationships or those embodied in prepositions and conjunctions. But how are we to know?

One thing is clear, however. Elements of language are not simply associations, such as conditioned reflexes. If language means relating in general, then the capacity for language means the capacity for certain relational activities. The activities cannot fail to be of a neurophysiological nature.

[4] This idea is suggested by the following train of thought: an organism could know nothing about the world if the world that surrounds it were ideally homogeneous–an ideal Ganzfeld for all senses, never changing over time. In this situation, there would be no contrasts of any kind, not even changes in the organism's own state that it could perceive (for otherwise there would not be total homogeneity for the sensory capacities of the organism). Any deviation from this non-plus-ultra of homogeneity, in either space or time, would at once produce differences that could be assessed by the organism as relations, and the relations would at once introduce notions such as *different from* (and, in man, also *here* and *there*, or *before* and *after*). This way of thinking goes to show that it is *plausible* to regard relations as the irreducible elements of knowledge in animals and man. (However, the thought is not a logical necessity; for instance, an organism could be perfectly tuned to certain types of energy and be responding to these and only these. Thus, if the world surrounding the organism were homogeneous but of property $x$ [homogeneously red, say], the organism would "know" something despite the homogeneity, namely redness.)

## APES: HEURISTICS OF CROSS-SPECIES COMPARISON OF COGNITIVE FUNCTION

How are we to know whether an organism, man or animal, has language knowledge? The most commonly used criteria are these: 1. the number of different words the subject produces; 2. the capacity to extend the meaning of the word to more than a single situation or object; 3. the capacity to emit an utterance in an "appropriate" situation; and 4. the propensity for combining words. Unfortunately, none of these criteria is rigorously enough defined to be of much weight in our judgment of language knowledge in a subject. Let us examine each one in detail. [Different criteria are, of course, possible. Hockett (1960) has used thirteen (for discussion, see Lenneberg, 1967).]

1. Let us first discuss the size of vocabulary. Recall what we have said about arithmetic. Clearly, we would not judge the degree of the subject's arithmetic knowledge by the size of his number vocabulary. The characteristic relationships inherent in arithmetic may be demonstrated in operations on groups of no more than four elements (or fewer, for that matter). The extension of the number vocabulary might be necessary to demonstrate that principles are being applied (relations computed) in producing answers to problems, instead of simple rote memory (if the system allowed only sixteen addition facts, it is conceivable that a subject would simply learn them by heart). But apart from this methodological consideration, nothing would be gained by increasing the group to more than four elements. If we wish to demonstrate a capacity for language, especially in an organism other than man, there is no particular need to insist on a large vocabulary. What matters is a demonstration that the organism has the capacity for performing the basic operations inherent in language, that is, of using words for classes defined by a set of relationships and for relating these classes to one another. Such a capacity can be demonstrated with fewer than 20 words; the size of the vocabulary does not tell us much.

2. The extension of meaning is, of course, an important aspect of the human use of words, but there are difficulties in evaluating this propensity, particularly when we are dealing with animals. Let me illustrate this. When my dog wishes to be let out, he gives a special little bark, while standing by the door. When I took him on a car trip recently, he suddenly gave the little bark. I stopped the car and opened the door, but the dog did not budge. I started to drive again and soon, the little bark. Again, no inclination to get out. At the next little bark I leaned over and opened the window a bit to let fresh air in; the dog immediately stuck out his nose and this was the end of the little barks. How shall I interpret this behavior? Shall I say "little bark means *open,*" and the animal has formed an intensive class based on the principle open? Or shall I say my dog barks in connection with discomfort? The first formulation would stamp the behavior as "language-like," the second not. There is no way of deciding between these

interpretations, because it is impossible to investigate the dog's mind or to know what he intended. Therefore, observations of this kind are essentially useless for judging language capacity. Because of the notion of tolerance with respect to the mapping of words onto relations, it is impossible to say when a subject has made language-like but *novel* (metaphorical) use of a word and when he has made the *wrong* use of a word due to lack of comprehension of the underlying semantic field. As mentioned before, language is different in this respect from arithmetic, where there is no tolerance for the particular mapping of a number word into the set of relations implied by the general concept of numbers. In arithmetic we can tell at once whether or not a word has been used correctly.

3. Now we come to that will-o-the-wisp, "appropriate situation." One of the peculiarities of natural languages (and of the use of natural languages) is that in one sense, there is no such thing as an appropriate physical situation for an utterance, and in another sense, any situation is appropriate. If a subject will say *x* only upon presentation of *x*, it is true that the situation is appropriate for *x*, but the use of the word may be quite odd. Characteristically, the sight of an object does not automatically elicit a verbal response in man. On the other hand, man uses language to communicate thoughts, dreams, desires, etc., and this occurs precisely in the absence of the object. But then again, there is nothing wrong with occasionally saying *x* upon seeing it. We see, therefore, that appropriateness of situation may be construed as either saying something in the absence or saying something in the presence of a given situation, and the final criterion for propriety is what we assume to be the intention of the speaker. Sometimes we can make educated guesses about the subject's mental state, but in many other situations this is not possible. Consider a subject who utters such words as "listen," or "one." There will never be a situation that is inappropriate to these words; though "listen" might customarily be uttered in the presence of loud noises, it could either be extended to soft noises or else be interpreted as meaning "lo! no noise." And whenever the subject says "one," there is one of something around. Suppose the subject has the word "food"; if the word is produced only at precisely the moment that it is "called for," we have no assurance that the word is understood in any sense other than "I am hungry," which is very "unlanguage-like." If it is used both at the appropriate moment and at other times, we may assume either that the general meaning of the word is not understood, or that it is understood and that the subject is telling us something about food at odd times. Human communications can take place in entirely appropriate situations but in the complete absence of language knowledge. An American sailor on liberty in Shanghai can pantomime his desire for food or female companionship by gestures that are appropriate to the situation, but the communication has nothing to do with Chinese, English, or any other natural language. Conversely, it is possible to tell a story that is understood by virtue of shared language knowledge but in the complete absence

of an appropriate situation. I think these examples show that appropriateness of situation is not a very satisfactory criterion for estimating the degree of language knowledge in a subject.[5]

4. Finally, let us examine the criterion of spontaneous combination of words. There is a temptation to regard any concatenation of elements as an illustration of the beginning of syntax. But the simple stringing up of meaningful units is not enough. *Tail-wagging* is a meaningful unit—at least we interpret it as meaning "my state is a happy one." *Looking-at* is a meaningful unit—at least we interpret it as meaning "the target of my gaze produces an internal response in me." If my dog looks at me and wags his tail, two units have been combined that might be freely translated as "Eric, I love you!" Concatenation, however, may be the result of random generation. Suppose a mathematician wishes to teach an animal a simple arithmetic language consisting of four symbols only. He finds that if he says "three," the animal responds "one, two." He interprets this as a primitive but essentially appropriate and correct "combination of words" by imputing to the animal the intention of wanting to say "one and two (is three)." Soon he discovers that when he says "two," the animal again says, "one, two," which delights him; now he has proved that the animal has learned multiplication. When the animal hears "one," it again says "one, two," which, of course, can mean nothing else but that it has caught on to subtraction. And when it says "one, two" upon presentation of "zero," his trainer finally has evidence that the animal has spontaneously learned to do addition with the residue class modulo three! The mathematician may, of course, be correct; but then again, the animal may be responding randomly. The combining of words is of interest only if we can demonstrate the particular type of relationship upon which the combination is based.

We have seen that of these four popular criteria, none is of much help in providing definitive evidence of language knowledge. The difficulty is due largely to the fact that saying something is not evidence of language comprehension. To test comprehension, it is necessary to examine the subject's understanding, either by giving commands that call for the combination of acts in specified ways or by asking questions that call for a yes or no answer. Probably it is easier to begin with commands. We must take care, however, that the commands are not simply the names of a small repertoire of tricks, as argued before (sit, speak, wave bye-bye, etc., will not do, because each elicits a unique response and need not involve any relationships; each is just the sign-stimulus for a response). Instead, we must use commands in which we test the comprehension of a small set of relationships. The paradigm is essentially that suggested first for

[5] The only sense in which "appropriate situation" may be used as an objective criterion is *syntactic;* in the "sentence," *Johnnie of wants a banana,* the word *of* does not occur in an appropriate situation. Appropriateness in this sense should be discussed under the heading of well-formedness in syntax. Whether intentionality may be excluded from such a discussion is still an open question.

arithmetic: Point to the number that equals 2 + 3. In other words, we determine the relationship between the numbers, instead of imputing the relationship to a pair of numbers uttered by the animal.

Let me give a concrete example of some feasible experiments. We shall see, however, that even positive results of this sort fail to prove definitively that chimpanzees can acquire a form of human language. At present, a chimpanzee, Washoe, is said to be learning to use "a form of human language" (the American Sign Language of the Deaf; Gardner & Gardner, 1969). So far, it is mostly the animal's achievement in word production that has been described. However, it should be easy to do some controlled experiments in which her understanding is also tested, and thus her comprehension demonstrated. We would continue to use sign language as a vehicle, and would confine ourselves essentially to the words that she is reported to have learned. In view of what we have said in the previous section, our experiments would have to concentrate on Washoe's capacity for understanding relationships, but her present vocabulary is rich enough for this.

Suggested procedure: We would use a double-blind test (following, for instance, Pfungst, 1911); the interpreter or observer of the animal responses must not know what the preceding commands have been.

Materials: The experiments would be conducted in a room full of objects, including some for which the animal has not yet learned any words. Further, there should be multiple instances of objects for which the animal does have names (several shoes, blankets, etc.).

Language: We would choose nine object words from her present vocabulary–*key, shoe, blanket, hat, bib, flower, toothbrush, pants, you*–and three relational words–*out* and *in,* plus the additional word *touch* (suggested because Washoe already uses the American Sign Language symbol for *touch* in the sense of tickle).

The first thing the animal would have to learn is that every command calls for a specific type of action. Examples: To the signal "bib in hat," Washoe would have to respond by putting one of the bibs inside one of the hats. The correct response to "key touch shoe" would be to place a key so that it was in contact with a shoe. The word *out* would, of course, be used as the inverse of *in.* If we included commands such as "shoe touch shoe" or "hat in hat," quite a variety of commands could be given. Half of the possible commands would be used for training Washoe; the other half would test her comprehension. (Since all of the experimenter's utterances would be understood as commands, no special sign for "do" would be needed.)

This language is quite simple, although it is impossible to say whether it is primitive in any "natural" sense. If these relationships could not be understood by the animal, perhaps some others could, and we might use our measure of success as a criterion for simplicity. However, unless some comprehension of relationships that exist in language were demonstrated, we could not even speak of "primitive" beginnings of language capacity.

If the animal were successful, we could complicate our first language model in two ways: we could complicate the syntax, and we could add to the repertoire of relations. Let us begin with the first. For instance, we could string up relationships, such as in the command: "key touch shoe in blanket." Suppose we also complicated the language in the other direction, and added the relationship *and.* This would allow us to test for an important feature in syntax and one that is of particular interest because of its perfect parallel in arithmetic. Consider the command "shoe and key in blanket." This is understood, at least by English speakers, as "shoe in blanket and key in blanket," but not as "shoe and key in shoe and blanket," which would be a possible interpretation, since there would be several instances of shoes and blankets present in the testing room. We can represent this in formal notation as follows:

Let "and" be represented as +, and let "in" be represented by *, then

$$A + B * C$$

is understood as

$$(A + B) * C = A * C + B * C$$

but not as

$$A + (B * C) \textit{ or } A + B * A + C.$$

In other words, it strikes us (as speakers of English) as logically obvious that the *in* relationship applies to both shoe and key, but that the *and* relationship does not apply to both key and blanket.

All languages have some notion of negation, and therefore also of affirmation. Suppose we introduced this concept now, so that negation could be used either to cancel the command as a whole or to cancel one of the relationships in a composition containing two relationships, such as "shoe and key in blanket." The reader may readily convince himself that this new concept makes it necessary to introduce a rather complex set of syntactic rules before it can be used in sentences unambiguously. Without syntactic rules to distinguish between commands such as "shoe and not key in blanket," "not key and shoe in blanket," "key and shoe not in blanket," etc., we would be unable to give unequivocal interpretations to these compositions. Children can answer questions by "yes" or "no," so that the basic concept of affirmation and negation in itself does not appear to be insuperably difficult; nevertheless, they ordinarily have a difficult time in learning the very complex syntax of negative constructions in English (Bellugi, 1971). In other words, the syntax of even the most primitive languages quickly becomes complicated by introducing more than just a few common relationships.

If we could introduce hand-signs for affirmation and negation, we would have prepared the situation for asking questions that demand a yes or no answer instead of an action. Every utterance could now be prefaced by a sign that tells whether it is a command ([Put] key in shoe) or whether it is a question requiring a "yes" or "no" answer ([Is] key in shoe?). This kind of exchange—question and affirmative or negative answer—is entirely universal among all languages of the world, and even small children can quite readily learn to answer simple questions. Could Washoe learn it?

These few examples of syntax should make it obvious that a natural language is more than an accumulation of items. The relations in use interact—relate to each other in peculiar ways—and thus the complexity of the relational structure as a whole increases exponentially with the addition of every new relation.

In addition to Washoe, another chimpanzee, Sarah, is being trained in language-like tasks by Premack (1969). This animal is reported to be composing sentences and answering questions of an absolutely amazing relational complexity. The vehicle of communication in this case is plastic chips of different shape and color, each standing for a word in English. The chips are taken out of a box and lined up on a board to form sentences. There is some more direct evidence, in the case of this animal, of comprehension of relations; unfortunately there are no films or slides available of the ape's spectacular success, and no double-blind testing with several independent observers has yet been reported.

While I admire the patience, the skill, the ingenuity, and often the courage behind these training endeavors, I would still have serious reservations with respect to their significance, even if all the experiments reported and proposed could be confirmed and extended. Neither the Gardners nor Premack claim (or expect to show) that there is *no difference* of any kind between the language acquisition capacity of chimpanzees and that of a child without a tongue or one that is congenitally deaf. Thus the species-specificity of the human form of language interaction and capacity (regardless of its vehicle, i.e., whether audio-lingual or visuotactual) remains uncontested. The work on primates can show only that *some* common denominators may exist, a circumstance that is impossible to deny even in the absence of the present training demonstrations. Let us suppose for a moment that the experiments proposed in this article had already been successfully concluded. I would still be hard pressed to say what they go to show. Should we see in these demonstrations a "primitive" form of man's capacity? The regular onset of human language development but arrested at an early stage? A similar but different form of language capacity? A homolog of our behavior? None of these formulations appears to be entirely justified (Lenneberg, 1969).

## BRAINS: PRELIMINARIES TO THEORIZING

Let us return to two earlier points. 1. No one can deny that man has, at the very least, an obviously different propensity for language than do other animal forms. 2. Relating something to something else is necessarily associated with a process, and if language is a set of relational operations, there must exist correlated neurophysiological processes. These two points lead us to ask a number of questions about the brain.

The conjecture that man's language capacity is based on biological species-specific peculiarities does not commit us to any particular hypothesis about anatomical or physiological brain correlates for language, and since so little is known about brains in general, I have, in the past, refrained from making any specific claims. This caused my critics to accuse me of an "extreme holistic" viewpoint (Whitaker, 1969) or of a failure to read the "obvious biological sign posts" (Guttman, 1968), which to them apparently spell out a clear story. I think it is much too early to state with confidence what it is in the brain that gives man his peculiar endowment for language. All we can do at present is to infer from the nature of language what general sorts of things we might look for in the brain–what kinds of information about the brain may or may not be relevant to an explanation of language capacity. I shall confine myself here to a critique of some of the more popular assumptions concerning brain correlates of language.

Suppose we could show by careful anatomic dissection of the human brain that the architecture of its connecting fibers is decidedly different from that of other primates and lower mammals. This would perhaps suggest that man's language capacity is due to a unique neuroanatomical specialization; but to lend credence to this, we would also have to make a strong case for the relevance of these peculiar fiber connections to that which we have characterized as the essential nature of language. If we were successful in the latter, we would have fairly well accounted for the mystery of why man has language but chimpanzees do not, and experiments such as those conducted on Washoe and Sarah would be superfluous. However, I think our insights from the first section can be shown to cast serious doubts on the relevance of fiber architecture to language capacities.

In general, we shall see that by combining the question "What brain correlates would we expect for arithmetic knowledge?" with the question "What brain correlates would we expect for language knowledge?" and by trying to answer them jointly, we can at least get a feeling for what is reasonable to expect and what is unreasonable. I shall use this approach to defend the following claims: (1) the architecture of intersensory fiber connections does not explain the capacity for either arithmetic or language operations; and (2) there are no cogent reasons to believe that language or arithmetic knowledge can be further subdivided in such a way that their components or attributes might be correlated with separate, autonomous neural mechanisms, each with a distinct cerebral locus.

1. *Fiber connections.* The concept of a number is independent of the physical shape of its symbol. One hundred may be represented as *100,* or by the

spoken word *satem,* or by the instruction "touch-all-fingers-on-both-hands-as-often-as-there-are-fingers-on-both-hands." There is no fixed association between any visual, auditory, or tactual pattern and the concept. Consequently we would not expect the basic capacity for arithmetic to be dependent upon any particular sensory input, and we are not surprised to hear that congenital deafness or blindness (or both) leaves the capacity for doing arithmetic unimpaired. If touch alone (in the deaf and blind) is sufficient to develop arithmetic capacity, it could hardly be argued that arithmetic is possible in man because of some felicitous fiber connection between auditory and visual projection areas.

That the situation for language capacity is quite similar to that for arithmetic in this respect may be seen from the clinical parallels. Neither deafness nor blindness nor the congenital presence of both bars the development of language through the vehicle of touch. There is no question but that an individual like Helen Keller could pass a language test of considerably greater complexity than that previously outlined. The argument of the preceding paragraph is directly applicable to language capacities. Relations such as *in, out, to, and, not, is* are not dependent on any sensory quality, and the same must be true of those relations that serve to define classes, such as furniture, freedom, grandmother, etc.; otherwise, it would be impossible to understand how a person who is both deaf and blind from birth can learn the meaning of these words. (There are a few exceptions: the meaning of some words does require specific patent sensory pathways; examples are words for color, for noise, for touch qualities, etc. However, it is curious to see how relatively unimportant is the incomprehension of such words in the language of the deaf or blind.)

Once a language has been established through whatever sensory avenue, it may continue to function even under conditions of dramatic sensory deprivation. Nor is it necessary for language operations that specific outside stimuli impinge on the speaker in order that a specific utterance may occur. Thus, it becomes difficult to maintain that anatomical connections between cortical projection areas can *explain* man's language capacity, or that a chimpanzee's difficulties in this respect are *due to an absence* of connections from, to, or between the primary visual and auditory projection areas of the cortex.

Furthermore, if we put too much faith in the importance of gross connections, we commit ourselves (by implication) to postulating rather fixed "centers of activities." There is evidence, however, that "activity centers" are not fixed for life but allow of some degree of shifting during the formative years of the brain. For example, the lateralization of language to the left is dependent upon an intact corpus callosum and the absence of pathology in the left. Patients with congenital absence of the corpus callosum do not seem to lateralize, and an epileptogenic focus on the left may prevent left lateralization altogether. Also, early cortical lesions, both in man and in other mammals, alter the course of normal functional differentiation of cerebral tissues and may, as a consequence, cause a somewhat different distribution of functional loci over the cortex. This relative plasticity in localization of function imposes severe strains on the logic of a strict connectivist explanation of our language capacity. I must stress here, that I am not ruling out the possibility that *connectivity* between cells is of a peculiar kind in man, constituting one of the prerequisites for our characteristic intellectual capacities, including language. But by *connectivity* I mean the fine structure of dendrites and axons. The relations expressed in language are undoubtedly dependent upon nervous activities, and these may well depend on the specific microscopic architecture of neuronal units. This is different from searching for gross subcortical or tangential fiber tracts.

2. *Indivisibility of language knowledge.* It must be clearly understood that my comments here do not refer to the general skills with which a person communicates verbally. I am concerned only with an alleged possibility of explaining the language capacities of man by gross anatomy. I do not deny that one can correlate clinical speech or language disorders with types and locations of lesions. However, it is not at all clear yet what the cause of this correlation is.

The primary prerequisite for human verbal communication is the capacity for language knowledge; in addition, several other skills are integrated with this capacity. It is possible to knock out some of these other skills by pathology without affecting language knowledge, and the capacity for language knowledge itself may also become disordered independently. In the clinic we see a great number of highly characteristic syndromes, each with its characteristic lesion at specific sites. Thus, from a clinical point of view, I would call myself a "localizer." Since it is certain that there are some brain lesions that leave language entirely intact, we may assert that there is a specific language territory, whatever its topography may be. The following discussion deals only with this territory and, more specifically, with the question whether it is possible to subdivide that territory further in such a way that individual language (knowledge) traits might be assigned to anatomical loci within this territory. The assertion that the latter is possible I call the "strict localization theory of language," and my aim is to disprove it.

Imagine some mechanism, say a computer; we ask the electrician who assembled the machine to give us a complete list of all of its components, and the user of the machine to give us a list of its behavioral traits. Our two consultants work independently of each other; neither knows what the other does. Now, it should be obvious that the items of the two lists will not necessarily stand in a one-to-one relation. Although there will be some items on the engineer's list that may be functionally related to individual behavioral traits (such that specific dysfunction here will produce specific dysfunction there),

there are, logically, four other possibilities, or altogether five ways in which the lists may be related:

| | | |
|---|---|---|
| (1) | correspondence is | one to one |
| (2) | | one to many |
| (3) | | many to one |
| (4) | | many to many |
| (5a) | a component: | one to none |
| (5b) | a trait: | one to none |

The first four are too obvious to need elaboration. An example of 5a would be a superfluous or redundant component; for instance, elimination or dysfunction in this case would have no behavioral consequences. In 5b, the list of behavioral traits might include an item actually not attributable to the mechanism of the machine at all, but due to some environmental conditions to which the machine corresponds indirectly; or the compiler of the behavioral traits might have failed to distinguish between behavioral principles (e.g., capacity for multiplication in general) and specific computations (frequent multiplication of $a \times b$).

I assume that a "strict localization theory" of language is one that expects perfect one-to-one correspondences between anatomically identifiable components of the cerebral language territory and specific traits of language. Such a theory makes two untenable assumptions.

The first is that one can compile a list of all the traits that are necessary and sufficient for the characterization of language. I doubt that this is either possible or relevant to the anatomy of the brain. Translate the argument to arithmetic. If we were to consider "any form of arithmetic" in general, our list would be simple: it would contain the single trait *relating.* This is unsatisfactory, because by this definition every organism has a primitive form of arithmetic. If we are to consider a specific arithmetic, such as that taught in grammar school, we would list particular relations (or axioms), and I do not believe anyone could hope to find cerebral loci that bear a unique functional relationship to these "components of arithmetic knowledge." In the case of language knowledge, the list of traits would be very long, and the particular entries would probably (at least for the time being) be rather arbitrary because of our difficulty in stating what the universal, common denominators of all possible relationships in languages are.

The second assumption is that traits can be analyzed as if they consisted of autonomous and independent entities. On the behavioral side, it is not clear whether such notions as processing, iteration, phoneme analyzing, and-relations, transformation rules, etc., are autonomous, separable traits. Their individuality may be challenged even on the grounds of logic, let alone on the grounds of psychological and physiological reality. If we cannot make sharp distinctions between these theoretical notions (which I believe can be shown to be true), how much less can we expect to find specific sites for them in the brain!

In summary, I do not deny (in fact, I firmly believe) that specific parts of the brain contribute in their own characteristic ways to physiological functions as a whole. But this does not mean that every aspect of physiological function has an obvious, unique behavioral correlate. Further, there is no question that verbal communication may be interfered with from specific sites and in characteristic ways. It is precisely findings of this sort that constitute strong evidence that our capacity for verbal behavior has a biological foundation. The fact that certain lesions produce characteristic clinical pictures of speech abnormality does not imply, however, an aggregate of independent traits, any one of which could be neatly abolished by a lesion. The advance of a wave of tumbling domino pieces may be arrested by the elimination of any two contiguous pieces; if a couple of particular pieces are particularly prone to elimination (because they happen to be close to a teenager's elbow) we must not attribute to just these pieces the exclusive function of wave advancement. Language operations may be thought of as a family of interrelated activity patterns, which may be deformed by local interference.

## DISCUSSION

Man (and probably many another species) appears to react primarily to *relative* differences in energy levels or, in more general terms, to be attuned to relations rather than to absolutes. There is evidence that species differ from one another in the types of relational configurations to which they are most sensitive. It is possible to discover empirically the hierarchy of pattern salience (or stimulus effectiveness) for any given species. If this is done for man, we would go from relations that seem to us "utterly obvious" to circumstances that seem "totally incomprehensible." In the case of man (and probably certain other animals, too), the essential relations that are reacted to are often quite independent of particular sensory qualities. Hence many patterns and configurations are recognized as similar, even though, for instance, one may be perceived tactually on the skin of the leg and the other auditorily (e.g., the pattern of one long and two short stimuli). Similarly, the relational patterns of language can be extracted from qualitatively different vehicles (speech sounds, Morse Code, graphic representations, tactual encoding, etc.). The same is true of arithmetic. Since it is the invariance in relational patterns that counts, while the sensory modality by which the stimulus is received is quite immaterial, one gets the impression that the extraction of particular relations from the environment is an *activity,* not a passive *receptivity.* The same idea is expressed by Piaget's notion of *construction* or by the more recent terms *computations* or *operations.* In each case, the words express the opposite of *simple storing, taking in, replicating.* Man (and, perhaps, chimpanzees) has a propensity for applying, to relations themselves, the relational computations used for processing sensory information; that is, man operates on the products of his own intellectual activities. Thus man does not only see or hear *two stimuli,* but can also count two *twos*; he does not name only classes of objects, but can also name names. Thus, conceptual systems are built up, of which language and arithmetic are but two examples. All of these systems have six important properties:

1. They are self-generating, in the sense that there can be numbers of numbers, names of names, relations of relations, pictures of pictures, thoughts of thoughts.
2. They consist of open sets or intensive classes, because the organism deals only with general characteristics.
3. The constructed systems consist of nothing but relations.
4. The relations constituting the system are readily represented by artifactual symbols.
5. The exact physical configuration of the symbols is of small concern to the symbol maker.
6. There are no other limitations to the quantity or quality of relations that make up a symbolic system than the biological constraints (the physiological limits to the operating characteristics) of the mechanism that computes the relations, that is, of the brain.

The argument of this article is best illustrated by means of a metaphor. The patterns of a kaleidoscope are infinitely variable; yet the range of this variability is delimited by the physical properties of the scope. No configuration can escape the limits. Thus each pattern that is formed has its unique structural characteristics, yet every pattern is related to every other. It is possible to take a single pattern and describe it, for instance by describing the position and nature of each piece. However, the gestalt that is formed is not a simple aggregate of these descriptive items; removing certain pieces would change the structural characteristics of the gestalt in more complex ways than simple subtraction. In this metaphor, the kaleidoscope corresponds to the brain; each pattern to one conceptual system; the constituents of the patterns to relations. I have pointed out (p. 163) that one can associate with every relation an activity or process (the relations that a computer can compute are dependent upon given activities that the machine can engage in). Thus the configurations we are concerned with in this article are configurations of activities or, in other words, the conceptual systems are *activity patterns.* We have shown how relations (as expressed, e.g., in rules) interact with one another, forming a system that is more than the simple sum of its components. In the case of the kaleidoscope, the patterns are end-products; they don't "do anything." In the case of the brain, however, the activity patterns that correspond to conceptual systems have an output: they produce symbols and classes of symbols, such as languages or algebras.

Let us pursue this metaphor a bit further. Suppose we want to find out how kaleidoscopes in general work, but have access to nothing but the patterns they make. Let our first task be to discover which patterns are made by the same kaleidoscope; next we hope to be able to group kaleidoscopes according to the similarity of their constitution. The first task has been achieved if we can at least specify a *class* of patterns by means of a *set* of transformations that can map any one pattern in the class into any other pattern in the same class. In this paper we have looked at two patterns, arithmetic and language, and the thesis has been proposed that one can map one into the other because they have both been generated by the same instrument, the human brain. Other, similar transformations might carry spoken English into the sign language of the deaf, or the logic underlying physics into the logic underlying chemistry. We still know nothing about the nature of the kaleidoscope.

Our next task is much more difficult. How can we arrange or order kaleidoscopes if we know nothing about their internal structure and function? In the case of brains of different animals, we know a bit about their anatomy—enough to know that each species has a species-specific morphology—almost nothing about how they make the patterns we are discussing, and very little about their most intimate, i.e., molecular, structure. We must, therefore, rely largely on the generated patterns for source material. We are aided by the obvious fact that brains have an ontogenetic history. The conceptual systems produced by an immature brain may have no formal equivalence to the systems of mature brains, and therefore, in terms of our metaphor, the immature and the mature brain each produces a distinct class of patterns with no overlap between them. Consequently, the set of transformations that defines one class is also distinct from the set of transformations that defines the other. Consider now that the set of transformations is actually a formalized description of the kaleidoscope. (It is not a blue-print of how the kaleidoscope works, but the "name" of the kaleidoscope in a formal metalanguage.) Given the fact that immature brains develop into mature brains, there must also exist a historical continuity from one set of transformations to the other giving us a formal criterion for arranging at least certain kaleidoscopes, or brains, into a single order; we might call this order a maturity sequence. (Immaturity is often equated with primitivity or simplicity; I find these words somewhat misleading, because they seem to imply more than is justified.)

Now we must try to arrange all the maturity sequences into a still further, higher order. The motivation for this is again biology. We have good reasons to believe that animals have a common origin and that their structural similarity reflects their proximity of descent. Because of this, we are convinced that there must be a yet more general, more abstract level on which one may construct transformations that relate the maturity sequences to one another. Now it is obvious that transformational pathways on this level should be much more involved than those on the previous level, because immature brains grow directly into mature brains, and therefore there is also historical continuity between their respective classes of patterns of conceptual systems; but the brains of chimpanzees do not grow directly into those of humans, and therefore any kind of construct that relates maturity sequences to one another is bound to be both more speculative and more arbitrary.

When chimpanzees are taught a form of human language, the implicit hope is to demonstrate phylogenetic continuity between their language capacity and ours. However, to show that some similarities exist tells us very little. If my thesis is correct that language is but one symbolic system out of a whole (intensive) class of such possible systems, all delimited by the instrument that generates the systems (the human brain), then the data from the chimpanzee "language experiments" is simply not rich enough to draw far-reaching conclusions about the proximity of chimpanzee language to human language. Nor are the results sufficient to construct theoretical pathways that relate chimpanzee capacities to human capacities. At the very least, chimpanzee language capacities must first be related to other conceptual achievements by these animals, so that a coherent picture emerges of all of the animal's intellectual activities. Only after this has been achieved does it make sense to make comparisons between the chimpanzee's achievements and human achievements, keeping in mind the necessarily speculative nature of this endeavor.

## REFERENCES


Bellugi, U. *How children say no.* Cambridge, Mass.: M.I.T. Press, 1971, in press.

Benacerraf, P., & Putnam, H. (Eds.) *Philosophy of mathematics: Selected readings.* Englewood Cliffs, N. J.: Prentice-Hall, 1964.

Gardner, R. A., & Gardner, B. T. Teaching sign language to a chimpanzee. *Science,* 1969, vol. 165, 664-672.

Guttman, N. Review of E. H. Lenneberg, *Biological foundations of language. Journal of the Acoustical Society of America,* 1968, vol. 43, 178-179.

Hockett, C. F. Logical considerations in the study of animal communication. In W. E. Lanyon & W. N. Tavolga (Eds.), *Animal communication.* Washington, D. C.: American Institute of Biological Sciences, 1960. Pp. 392-430.

Lenneberg, E. H. *Biological foundations of language.* New York: Wiley, 1967.

Lenneberg, E. H. On explaining language. *Science,* 1969, vol. 164, 635-643.

Lenneberg, E. H. Brain correlates of language. In F. O. Schmitt (Ed.), *The neurosciences: Second study program.* New York: Rockefeller University Press, 1970.

Pfungst, O. *Clever Hans, the horse of Mr. Von Osten.* New York: Holt, 1911.

Premack, D. A functional analysis of language. Invited address, Division I, American Psychological Association, Washington, D. C., 1969. Mimeographed pre-publication tion copy.

Whitaker, H. A. On the representation of language in the brain. *UCLA Working Papers in Phonetics,* 1969, No. 12.

Zeeman, E. C., & Buneman, O. P. Tolerance spaces and the brain. In C. H. Waddington (Ed.), *Towards a theoretical biology; an I.U.B.S. symposium.* Chicago: Aldine, 1968.

## DISCUSSION

In psycholinguistics, a distinction is often made between "performance" and "competence," a participant said to Dr. Lenneberg. In dealing with chimpanzees, all you have to work with is behavior. If the animal fails to demonstrate language use, how can you determine whether this failure is due to inability to acquire the competence for language or language-like use, or to an insufficiency of the mechanisms necessary to make language performance adequate? For example, a language experiment might be so complicated that it would exceed the memory span of the animal or human subject. Failure to pass the test would not necessarily indicate that the subject lacked language knowledge.

Dr. Lenneberg has identified two variables that are important for human language, another participant said. One is a set of labels which are probably quite arbitrary. The other is a set of relationships which are probably not arbitrary. Are these relationships learned or innate? If they are learned, what is the primary or first-learned relationship? What is the mechanism in the brain which enables human beings to learn these relationships?

The controversy over the distinction between "learned" and "innate" or between "performance" and "competence" seems to me very tiresome and unproductive, Dr. Lenneberg replied. What do these words mean? It is hard to take knowledge apart in such a way as to be left with totally "learned" or totally "unlearned" components. Take arithmetic: it is the structure as a whole that constitutes "arithmetic knowledge." At the present time, it is virtually impossible to prove or disprove what aspects of language knowledge an individual has, how it came about, or whether his form of knowledge is identical to that of someone else (particularly since language knowledge does not necessarily require verbalness). It is perfectly conceivable to have a system that has all the formal characteristics of natural language without using any words at all.

As for arbitrary labeling and nonarbitrary relationships, neither arbitrariness nor the existence of several relationships tells us much about the characteristics of language, Dr. Lenneberg continued. Neither does the joining of units, considered by itself. Arbitrary associations as well as concatenation of elements are found throughout the animal kingdom. The joining of units is pertinent to language only if it indicates a certain (semantic) relationship; a random joining is a relationship also, but it is a meaningless one, the wrong relationship for language. In an experiment this can be controlled by the experimenter who dictates what the relationship shall be. If he says, "eraser on watch," and the subject puts the eraser on the watch, it has been shown that the subject understands and, therefore, that the subject has knowledge of the relationship "on." Or if the experimenter asks, "Is the eraser on the watch?" and the subject gives his signal for yes, he is giving evidence of understanding the relationships.

As for what the primary or fundamental relationships in language are, there seems to be no way to derive a primary order. Efforts to look at child language or to attempt to find common denominators in all natural languages have been made but with uncertain results. Thus far, in studies of child language, too much attention has been devoted to what children *say*, instead of what they *know*. What people say is not the same as what they know. For example, children come to speech clinics who do not use language at all to communicate, yet they can repeat long television commercials verbatim, with every intonation and detail so accurate you can even recognize the announcer's voice. Some people say such children do not have language; others say they do, but simply make very odd use of it. But no one really knows whether language knowledge is present in these children. We can only guess at what is actually present inside their brains, or whether it is the same knowledge that other people have.

You have suggested that some of the criteria people use to evaluate the degree of language knowledge that is present are inappropriate, a participant said. If these definitions and criteria are wrong or imprecise, what are the implications for teaching language to children? If there is a sequence of grammatical or syntactical construction, should we be displaying more of one kind than another at a particular point?

This comes back to the question of primacy, Dr. Lenneberg replied. If we could break language down in terms of what relationships develop first, we would have a marvelous tool for teaching children. But we cannot establish this convenient, orderly sort of hierarchy, even though the way children begin to use language does seem to have certain regularities, and be somewhat predictable. There probably is some sort of learning order, but until we understand what it is, the main thing we need to do is to *talk* to children. If the child does not understand a relationship when it is expressed one way, all we can do is try to find a way to make it more explicit, which we may or may not be able to do. There are many ways of interpreting a given situation, however, and this often results in inconsistent verbal behavior.

Teaching language to deaf children is an interesting problem; it is not very efficient as it is most frequently done, Dr. Lenneberg continued. Few of the children end up knowing and using language the same way that nondeaf people do. Usually, the deaf child is introduced to language through writing, at about age six, and the teacher tries to teach him grammar at that time. For example, the deaf child might be asked to say what happened last night and so he writes on the board, "Daddy TV." The teacher then writes on the board, "Verb? Where is the verb?" but the child has no idea what a verb is. Eventually he gets the idea that he should add the word "watched," and the teacher thinks, "Lo and behold, the child knows grammar." But it usually turns out this is not true; the child continues to make the same kinds of errors over and over, even though he may also grasp some very subtle language distinctions, often of the kind that the teacher has made no conscious effort to teach. Too much grammar instruction may confuse the child more than it helps him.

We might understand the nature of language knowledge better if we could answer the important question of whether animals establish relationships in the same way humans do. It is possible that the human brain is able to compute some types of relationships that given animals cannot, but this has not been demonstrated. Certainly, some properties of language are common to all primates, but others appear to be specific to man. This does not mean that man has nothing in common with other animals. For example, the relationships we find in language have something to do with cognition. But human cognition appears to be a special case of a more general type of cognition also found in primates, and, perhaps, other mammals. Although species do not shade into each other and each species has had its own history, this does not rule out some of them having certain traits and capacities in common.

You have spoken primarily in terms of the internal activity of the computerlike brain in bringing out language relationships, a participant said. Is there any aspect of language that has to do with seeking the information or selecting the inputs to be used by the computer in establishing these relationships?

Answers to questions of this sort tend to be philosophical rather than factual, Dr. Lenneberg said. After all, there is no little man who screens incoming information for language use. Selective attention is an unsolved problem as far as neurobiology is concerned.

# THE INTERPERSONAL REGULATION OF BEHAVIOR

*Robert M. Krauss*
Columbia University

To put the matter baldly, my feelings about interdisciplinary undertakings tend to be somewhat negative. Too often in the past such conferences have reminded me of Oscar Wilde's understandably cynical view of marriage: "an alliance entered into by men and women who were either tired or curious and who, in both cases, were bound to be disappointed."

For a psychologist interested in language, such a sentiment seems particularly perverse. For of all the areas of psychology, few have benefited as much as has the psychology of language by large infusions of knowledge from a related discipline, namely linguistics. Indeed, I think it is not unfair to say that the traditional psychological view of language behavior, and more specifically of the acquisition of language, has been radically changed as a result of the findings of modern linguistic science. This is familiar ground to many of you and I will not take the time to cover it here.

The impact of linguistics has led psychologists to focus their concerns on certain aspects of language use and to devote relatively less attention to others. The general problem that linguistic theory poses for psychologists centers about the notion of "competence"—the complex, abstract knowledge that underlies all language behavior and which a user must possess to comprehend and produce sentences. Linguists have been able convincingly to demonstrate the unforeseen complexity of adult linguistic competence. As a consequence, psychologists interested in language development have been led to investigate the process by which the child extracts "... from a finite sample of speech to which he is exposed the latent structure that will generate an infinite set of sentences [Brown, 1968, p. 49]."

But competence in the use of language develops for a purpose and to a social psychologist (which I confess to be) the main function of language is in communication. The words we utter affect the behavior of others and, indeed, mediate the greater part of social life. Viewed in this context, language use presents some problems that seem particularly psychological and for which our colleagues in linguistics have provided no ready-made solutions. Let me illustrate one of these problems by describing briefly an experiment (Krauss, Vivekananthan, & Weinheimer, 1968).

A large number of female subjects were asked to name each of a set of 24 color chips. Half were instructed to name them so they themselves could identify the colors at some later date; the other half were instructed to name them so that some other girl could identify the colors. We call the second condition *social encoding* and the first *nonsocial encoding.* Approximately two weeks later, all subjects were called back and asked to match names with colors. One third of the names were the ones she had given two weeks earlier; another third were names given by another subject under social instructions; and the final third were given by yet another subject under nonsocial instructions. The names were, of course, randomly intermixed. Subjects were best at identifying colors from names that they themselves had given; they were next best using names given by another subject under social encoding instructions; and they were worst trying to identify colors from names given by another subject under nonsocial encoding instructions. In addition, there were significant lexical differences between the names given under social and nonsocial instructions.

Now the point of interest about this experiment is that subjects in the two experimental naming conditions were talking about the same ostensible referents. That is to say, the referents in the two conditions were physically the same stimuli. What varied was the eventual target of their messages—the subject herself or some other person—and it was this fact that produced the differences in the messages. Incidentally, we have recently replicated this experiment using pictures of faces, rather than colors, with identical results.

Now any message, even the simplest, can be encoded in a number of alternative ways and at least some of the time a speaker's choice from among the various alternatives available to him will determine the listener's ability to comprehend the message. We can refer to a particular bird as *the bird, the ring dove, Streptopelia risoria,* or *birdie.* Which we choose will be determined in part by which of the terms we have available in our repertoire. But given that we have available more than one, our choice will also be determined by certain of the characteristics of our listener. In my reading of linguistic theory there is nothing that tells me how speakers make such choices—how one is able to infer from a set of assumptions about his listener the appropriateness of alternative ways of encoding a given informative content. As Brown has put it, "Effective encoding requires that the point of view of the auditor be realistically imagined [Brown, 1965, p. 132]" and both experimental research and everyday observation seem to confirm the existence of this process in adult speakers.

This linguistic fact generates a problem of real interest for any psychological account of the development of language use. For some time now it has been clear that the child's speech differs in at least one striking way from its adult counterpart. For an early observation and consideration of this difference we are indebted to Jean Piaget—the source of so much of interest and importance in the study of human cognitive development. I cannot improve upon Piaget's description. Writing in 1926, he commented:

> Although [the child] talks incessantly to his neighbors, he rarely places himself at their point of view. He speaks to them for the most part as if he were alone, and as if he were thinking aloud. He speaks, therefore, in a language which disregards the precise shade of meaning in things and ignores the particular angle from which they are viewed . . . In a word, the child hardly ever asks himself whether he has been understood. To him, that goes without saying, for he does not think of others when he talks [Piaget, 1926, p. 60].

Piaget's description of the rather ineffective attempts of young children to communicate, which has been confirmed by numerous investigators, stands in sharp contrast to the sorts of sophisticated communication skills employed by adults. And it was this discrepancy which led my colleague, Sam Glucksberg of Princeton University, and me to undertake a series of experiments over the last few years. I would like to describe some of these experiments and the results we have obtained and then go on to outline some of the cognitive processes which appear to us to underlie the use of language in communication.

Our general strategy has been to employ a standardized and highly schematic two-person communication task. Subjects are given the task of communicating about a set of novel graphic designs. The set of designs is illustrated in Figure 1. The property they have in common is their low codeability (Brown & Lenneberg, 1954). That is, they are difficult to name or characterize, each one eliciting from a group of speakers a wide variety of verbal labels. The designs are reproduced on the four vertical facets of a 2-inch wooden cube, one design on each. Each block has a hole drilled vertically through its center so that it can be stacked on a wooden dowel. The experimental situation is illustrated in Figure 2. There are two subjects, designated the speaker and the listener, each of whom is given a duplicate set of blocks imprinted with the novel designs. The speaker receives his blocks in a dispenser so constructed that the blocks can only be removed one at a time in a predetermined order. The listener receives the six

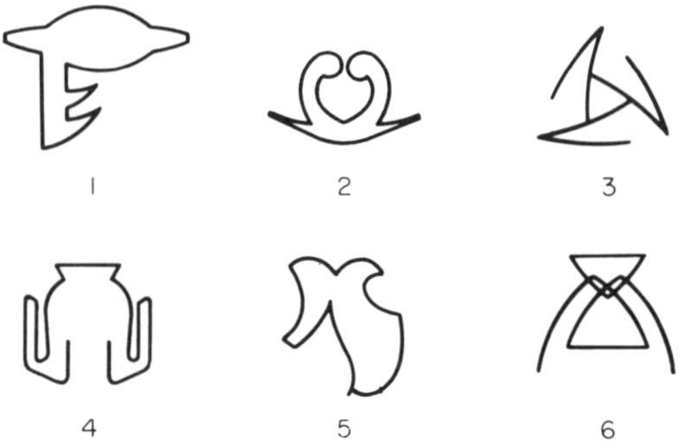


**Fig. 1.** The graphic designs employed.

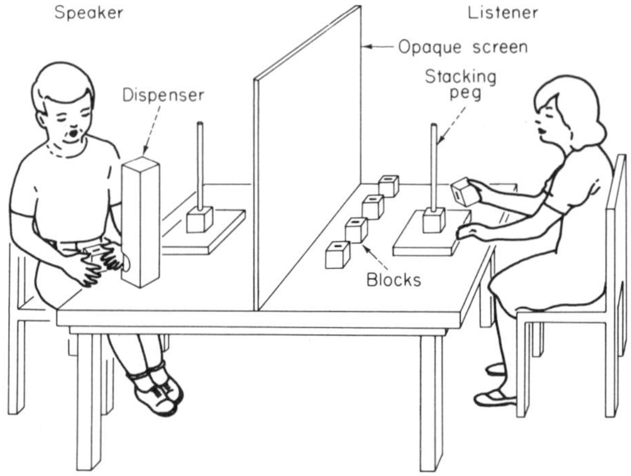


**Fig. 2.** Experimental situation. (Although a male speaker and female listener are shown, most of the studies employ same-sex pairs.)

blocks spread out before her in random order. The subjects are separated by an opaque screen so situated that they can neither see each other nor each other's blocks.

The task is introduced to the subjects as a game called "Stack the Blocks." The object of the game is to build two identical stacks of six blocks. The speaker is instructed to remove the blocks from the dispenser one at a time and stack them on his peg. At the same time, he is told, he must instruct his partner, the listener, which block to stack on her peg. No restrictions are placed on either subject's speech. In one set of experimental conditions, after the six blocks have been stacked the two stacks are compared in full view of both subjects.

Before playing the game with the novel designs, subjects are given several pretraining trials using a set of blocks imprinted with familiar objects (animals, circus figures, etc.). Since virtually all children can identify and name the figures depicted on the pretraining blocks, this procedure greatly simplifies teaching the rules of the game. At the same time it ensures that defective performance on the experimental task can be attributed to difficulties in dealing with the novel designs and not simply to an inability to follow the rules of the game.

In the initial pretraining trials subjects played the game in full view of one another. If after six trials they were unable to complete a single errorless trial, the experiment was terminated. Subjects who did complete an errorless trial in the face-to-face condition were seated on opposite sides of the opaque screen and given four more trials with the pretraining materials. Subjects who met a training criterion of two consecutive errorless trials proceeded, without further explanation, to play the game with the novel figures. Subjects who did not meet this pretraining criterion were discarded.

This, then, describes the basic task. In our research we have employed it along with several variations. For example, the experimenter can play the role of the speaker and "feed" the listener-subject with standardized messages of known informative value. Or names can be elicited from a speaker on one occasion and fed back to him or to some other person on another occasion.

In our first set of experiments (Glucksberg, Krauss, & Weisberg, 1966), we drew our subjects from a population of children ranging in age from 33–63 months in two private nursery schools in Princeton, New Jersey. Virtually all of the children were from middle class backgrounds and their parents were predominantly college educated, often beyond the undergraduate level. Conveniently, the schools were divided into two grades, a younger group with ages ranging from 33–49 months and an older group of 52–63 months.

Six randomly selected pairs from the younger group were run on the pretraining task. None were able to meet the performance criterion. For the most part these younger children were unable to follow the rules of the game using the familiar pretraining materials even when performing in full view of one another.

Accordingly, the novel designs were used only with the older group of subjects. Seven pairs of children in the 52–63-month-old group performed the experimental task after all had met the pretraining criterion. There were three male, three female, and one male-female pair. All pairs were run for eight consecutive trials. Not one pair was able to complete a single errorless trial using the experimental materials. Indeed, none showed any systematic decrease in the number of errors over trials. However, when the pretraining materials were reintroduced subjects performed without error, as they had previously. The subjects' difficulties, then, are attributable to the communication problem posed by the novel figures.

When we examined transcripts of these sessions, we were struck by the extent to which our subjects' messages were idiosyncratic and apparently unrelated to the stimulus figures. The imagery they employed seemed to be private, rather than socially shared or conventional. From observations of performance in Experiment I and on the basis of these and other theoretical considerations, two related questions were raised:

1. To what extent were children in our subjects' age range capable of performing adequately as listeners, given that the messages provided them are utterances of competent communicators?

2. To what extent can a message uttered by young speakers convey information to the person who uttered it, despite the fact that it apparently conveys little or no information to others?

The first question was studied in an experiment in which the experimenter played the role of speaker and provided standardized messages for a young subject-listener. Ideally, we would have liked to pair a naive, competent adult with each child, but in the settings in which the experiment had to be run this was simply infeasible. Instead, we selected names for the novel figures which had been uttered by adults in an earlier experiment (Krauss and Weinheimer, 1966) and which had elicited the correct response from an adult listener. These were pretested on an independent sample of naive adults and only names which elicited correct responses with certainty were retained.

Twelve nursery school students (age 46–63 months), drawn from the same population as in the previous experiment, were used as subjects. They first learned the listener's role on the pretraining task, with an experimenter playing the role of speaker. All twelve met the performance criterion. The novel figures were then introduced again with an experimenter acting as a speaker. Of the twelve subjects we ran, eight were able to meet a performance criterion of two consecutive correct trials within eight trials. These results indicate that a substantial proportion of our subjects are able to perform the listener role with some measure of adequacy, given competently formulated messages. They also suggest that to a large extent the inadequate performance of our speaker-listener dyads was attributable to deficiencies in the ability of our speakers to formulate messages that were socially adequate (i.e., were sensitive to the point of view of the listener).

The rationale for the experiment designed to investigate the second question deserves some explanation. It is clear that adult speakers distinguish sharply between messages formulated for oneself and messages formulated for others (see Krauss *et al.*, 1968; Werner & Kaplan, 1963). We have implicitly assumed that the difficulties experienced by our young subjects were due to their inability or disinclination to make such a distinction (see Flavell, 1968, for an informed discussion of this matter). Clearly such an inference is unwarranted from our data alone. Simply because our speakers' utterances were incomprehensible to others, there is no reason to believe they necessarily were meaningful to the speakers themselves. It conceivably could have been the case that these subjects simply assigned names to the figures in a blind fashion, without any meaningful connection between the names and the figures to which they were supposed to refer. To test this, six subjects in the 47–59 month age range (exhausting the available population of the nursery) were first asked to name the familiar figures used on the pretraining task. An experimenter then used these names in teaching the subject the communication task, again with the experimenter playing the role of the speaker and the subject as listener. When they successfully met the pretraining criterion, as all did, the novel figures were brought out and the subjects were asked to name these. One subject, who could not or would not name these figures, was dropped from the experiment. The remaining five subjects performed at a very high level of accuracy when communicated to by the experimenter using the names they had uttered. Indeed, on the very first trial only a single error was made. The names supplied by these children were quite adequate to guide their correct identification, although as in the experiment reported earlier these names were short and idiosyncratic. The entire set of names are shown in Table 1. Note, for example, names given by Subject 1 for Forms 4 and 5, *Daddy's shirt* and *another Daddy's shirt* (it is not clear whether *another* modifies *Daddy's* or *shirt*). Neither of these names, nor anything like them, appears in our adult protocols and the likelihood of confusing the two seems great. In addition, neither name seems particularly descriptive of the figures to which they refer. Yet this subject was able to make the correct selections without error.

TABLE 1
*Reference Phrases Given by Nursery School Children*

| | Subject | | | | |
|---|---|---|---|---|---|
| Form | 1 | 2 | 3 | 4 | 5 |
| 1 | Man's legs | Airplane | Drapeholder | Zebra | Flying saucer |
| 2 | Mother's hat | Ring | Keyhold | Lion | Snake |
| 3 | Somebody running | Eagle | Throwing sticks | Strip-stripe | Wire |
| 4 | Daddy's shirt | Milk jug | Shoe hold | Coffee | Dog |
| 5 | Another Daddy's shirt | Bird | Dress hold | Dress | Knife |
| 6 | Mother's dress | Ideal | Digger hold | Cater-pillar | Ghost |

These results indicated the advisability of broadening the age range of our sample to examine the sorts of developmental trends present. We did this in an experiment conducted at the Milton Hershey School in Hershey, Pennsylvania (Krauss & Glucksberg, 1969). Our subjects were 74 male students–14 in kindergarten, 18 in first grade, 20 in third grade, and 22 in fifth grade. (The unequal numbers reflect the availability of subjects in the various grades.) Subjects were randomly paired within grade and put through the pretraining procedure for the communication task described above. All met the pretraining criterion and went on to play the game with the novel figures. It was originally our intention to run all pairs for a total of 15 trials. However, the school's routine required that the entire experimental session (including pretraining) be completed in 50 minutes. For certain subjects, primarily among the younger ones, it was impossible to complete 15 trials in this time. In addition, even had more time been available it is questionable whether it could have been utilized since subjects who were unable to communicate effectively in several trials, again primarily among the younger ones, tended to become restless and uncooperative after 30–40 minutes. All pairs were run for at least eight trials, and only data from these eight trials will be considered.

The mean number of errors for grade levels is plotted over trials in Figure 3. It will be seen that initial performance differs little in the four age categories; however, the groups differ markedly in the rate at which they reduce errors. By trial 8; third and fifth graders are making fewer than one error on the average, while the initially poor performance of kindergartners shows no improvement whatever. First graders fall somewhere between these extremes. An analysis of variance of these data indicates significant overall differences between grades and over trials. A significant Grade X Trial interaction reflects differences in the slopes of the curves for the four groups.

These data demonstrate marked differences in communication effectiveness as a function of age, and our previous work suggested to us that such failures in communication were largely attributable to inadequacies in the speakers' encoding ability. However, this cannot be concluded from the data presented in Figure 3, since both speaker and listener were of approximately the same age. That is, it is possible (albeit unlikely) that speakers of different ages were actually equally good and that the observed differences are attributable to age-related variation in listeners' comprehension ability.

To test this we gave the messages our young speakers had uttered to a group of college students and asked them to identify the figure each message referred to. These messages were taken from the first trial of the previous experiment, because we felt that names uttered on the first trial would be less affected by characteristics of the listener than names used later in the session. In Figure 4 the mean number of correct identifications by adult respondents is plotted as a function of the speaker's grade. An analysis of variance reveals significant variation among these means.

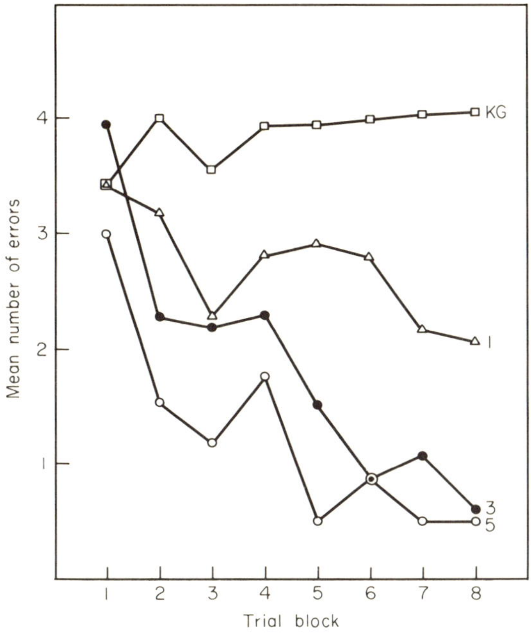


**Fig. 3.** Mean errors over trials for matched-age pairs in the four grades.

Note one interesting thing: adult performance is ordered as an increasing function of speaker's grade. However, on trial 1 of the previous experiment, from which the names used in this experiment were taken, no such ordering is observed. You may recall from Figure 3 that all groups started at approximately the same level. (The largest difference on trial 1, between grades three and five, was the only one coming anywhere close to being significant [$.10 > p > .05$ by *t* test] and the rest clearly were not.) Only on later trials did the differences between groups emerge. Clearly, messages uttered by speakers of different ages contain different amounts of useful information and this undoubtedly contributes to the age-related variation in the performance of speaker-listener dyads. But it is equally the case that the youngsters in the listener role were unable to utilize all of the information present. This result suggests that although listener proficiency may well develop earlier than speaker proficiency (Cohen & Klein, 1968; Glucksberg *et al.,* 1966), in neither ability do our grade school subjects approach adult competence.

One aspect of younger children's behavior that we had observed in these earlier studies involved the reaction to a listener's feedback concerning message understandability. We had hypothesized earlier that in nonsocial encoding the speaker's "name" for a figure is essentially unaffected by his listener's post-message utterances. This notion is consistent with Piaget's (1924) concept

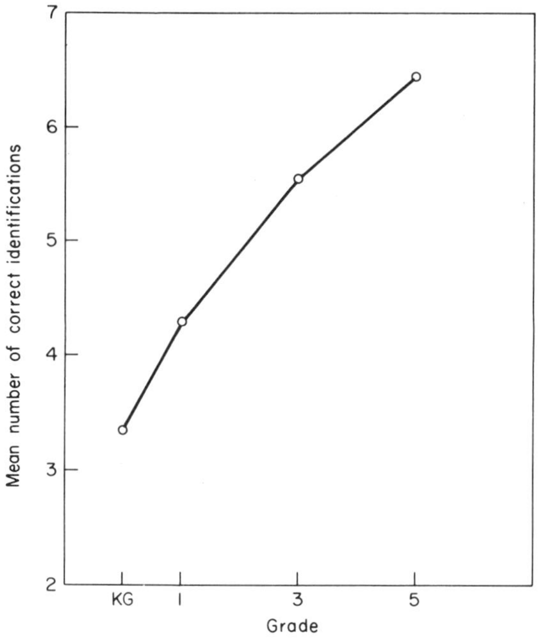


**Fig. 4.** Accuracy as a function of speaker's grade.

of nominalism: the tendency to treat a "name" as an integral attribute of an object, an attribute which is invariant and not subject to arbitrary change. With these ideas in mind, we conducted a study (Glucksberg & Krauss, 1967) to examine speakers' responses to listener feedback as a function of age.

Kindergarten, first-, third-, fifth-grade children, and college students, were assigned the role of speaker in our two-person communication task, with an adult experimenter playing the listener role. On the first trial with the designs, the experimenter said "OK," indicating understanding, after the description of the first, third, and sixth blocks. After the descriptions of the second, fourth and fifth blocks, the experimenter said to 12 of the subjects in each age group, "I don't understand which one you mean." To another 12 subjects in each age group he said, "Tell me more about it." And to a third group of 12 subjects he said, "I don't understand which one you mean; tell me more about it."

Since these three different messages made no difference in the behavior of *S*s, *S*s were pooled within age groups. The data indicate that younger children, in addition to displaying limited response repertoires, failed to edit., i.e., they did not modify their messages in socially appropriate ways. Figure 5 presents some of these data. Each bar represents the percentage of subjects in each age group, who, on at least one occasion, displayed one or more instances of behavior falling into each category. The first category, *New Description,* refers to a post-feedback description which likened the figure to a different object. For example, if the initial description was "it's like a boat," the new description might be "it looks like a hat." A *Modified Description* is one which preserves the major outlines of the original description but adds or changes detail. For the example noted above a modified description might be "a boat with a motor hanging down in back." In a *Repeated Description* the original description is repeated in substance, without modification. In a large number of cases such repeats were verbatim. The fourth category, *Silence,* is self-explanatory. From Figure 5 it is clear that socially appropriate behavior, as measured by these indices, increases with age. Kindergartners seldom give entirely new descriptions and they modify their initial descriptions considerably less often than do the older children. (Incidentally, this criterion is a relatively liberal one. It does not consider whether the modification was any improvement over the original; often it was not.) What kindergartners do with fair frequency is just what one should *not* do to communicate effectively: repeat the initial response. Note that this is never done by fifth graders or by our adult subjects. The data on "pointing" (not shown here) are particularly revealing. Only kindergartners and first graders "point"—that is, say things like "it goes like this" while tracing the design with a finger. The inappropriateness of such a response is particularly clear when one recalls that speaker and listener cannot see each other.

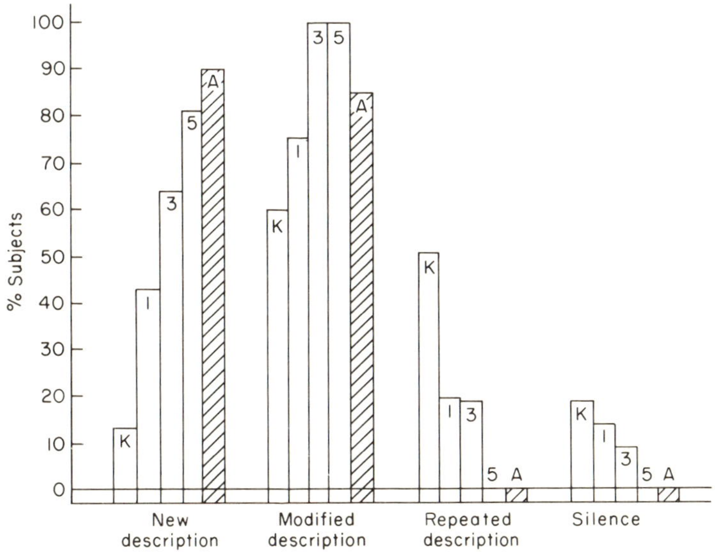


**Fig. 5.** Response to feedback as a function of grade.

In recent years, several models have been proposed for message transmission in referential communication (Cohen & Klein, 1968; Glucksberg & Krauss, 1967). Although these models differ rather markedly in specific technical detail—details that need not concern us for the moment—they are in agreement on certain general features. Each assumes the existence of a hypothetical repertoire of responses associated with (or, if you prefer, elicited by) any particular referent. Second, the models posit some mechanism by which responses are selected or sampled from such a repertoire as potential candidates for transmission in a message. Finally, these models interpose an editing or evaluation process between sampling and transmission in which the speaker attempts to evaluate the communicative adequacy of a tentatively selected response. Messages that fail to meet some criterion of adequacy are rejected, and the speaker iterates through the sampling and evaluation process until an adequately communicative response is obtained.

The precise nature of the mechanisms the various models posit are of less concern to the present discussion than is an implication they all hold for the potential sources of ineffective communication. Deficient communication can result from either of two reasons. First, a speaker may simply not have available in his repertoire responses that are adequately communicative. For example, Heider and Olivier in a recent unpublished paper demonstrated that subjects drawn from the Dani tribe, a primitive people in northern New Guinea (West Irian), were quite incapable of communicating about stimuli the criterial attributes of which were subtle differences in color. The Dani basic color vocabulary is quite meager. Essentially it consists of two terms—roughly, light and dark with appropriate modifiers. Heider and Olivier's subjects simply did not have available a vocabulary capable of encoding a precise shade of, say, red. Speakers of English, with its relatively rich color vocabulary, perform quite adequately at the same task.

But given that a speaker has available in his vocabulary an adequate response, there is no guarantee that he will select it. That is to say, his performance may be defective at the evaluation stage. He may select a term that is predictably not meaningful for a specific listener. For example he may, as one of our subjects did, call a stimulus figure "Mommy's hat," knowing full well that his listener had no conceivable way of knowing what "Mommy's hat" looks like. Or he may denote a referent he wants to communicate by a term that is equally applicable to the objects he wishes to distinguish it from. One of our subjects called one of our stimuli "a design"—a term that applies equally well to the particular figure he wanted to designate and to the other five figures in the set.

As a preliminary approach to understanding the sorts of cognitive processes which give rise to age-related differences in communication effectiveness, we performed the following analysis. In nearly all of our experiments we have transcribed our subjects' utterances verbatim. The data I will discuss are based on subjects in kindergarten, fifth, and eighth grade. We chose these transcripts for analysis because it seemed more likely that the wide age range would maximize the possibility of detecting differences.

What we did was to count the number of conceptual responses for each speaker's messages on the first trial. We chose the first trial because these messages are uncontaminated by contributions from the speaker's partner in a way that later messages may not be. By a conceptual response we mean the mention of a distinct descriptive element of a figure. So if a subject said that a figure looked like "snakes in a saucer" we would count two conceptual responses: snakes and saucer. The coding scheme is somewhat rough, but coders can use it with reasonably good reliability once they have some practice.

We would expect that the conceptual repertoires of our subjects would increase with age and that this would be reflected in the number of conceptual responses they gave in the communication task. And we are making the strong assumption here that the pooled responses of a particular age group is a fair estimate of the repertoire of any member of that population. This is certainly an arguable assumption, but there is at least some justification for it. The data are shown in Figure 6, which plots the mean number of conceptual responses for kindergartners, fifth, and eighth graders. Kindergartners on the average give only about two thirds as many conceptual responses as fifth graders. Fifth graders and eighth graders give roughly the same number of conceptual responses—if anything the margin is trivially in favor of the younger group. We were somewhat surprised by this result but the wisdom of hindsight convinces us that it should not be too troubling. But if we look at the communication accuracy data for these same subjects, one fact is quite obvious: eighth-grade speakers are more adequate communicators than fifth graders. The difference in number of errors for fifth- and eighth-grade speakers communicating with their age peers is clear. Consistently, over twelve repetitive trials, fifth graders make roughly twice as many errors as eighth graders. If one raises the objection that this does not take into account the effect of the listener's age (and we know from previous work that the listener's age does have an effect), we can compare the performance of these two groups of subjects talking to listeners of the same age—namely kindergartners. In such a comparison, the difference between fifth- and eighth-grade communication accuracy is accentuated. Kindergartners are at best woefully inadequate partners but their performance when paired with eighth graders is vastly superior to what they do with fifth graders. Indeed, the performance of kindergarten listeners with fifth-grade speakers shows relatively little improvement over trials.

Recall, however, that the conceptual repertoires of fifth- and eighth-grade speakers, to the extent that they are tapped by our measure, differ not at all. How, then can we explain the increased effectiveness of the older subjects? Returning to our earlier argument, we hypothesized that a large repertoire was a

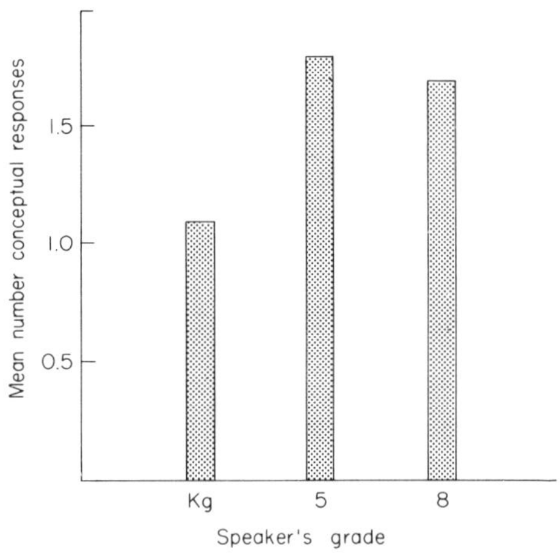


**Fig. 6.** Mean number of conceptual responses per figure.

necessary but insufficient condition for effective communication. In addition to the repertoire, it is necessary that a speaker have the ability to select out of that repertoire those concepts that are at least potentially socially meaningful.

Unfortunately the communicative meaningfulness of a given conceptual response, although intuitively reasonably clear, is difficult to characterize operationally. If one who is familiar with the six stimulus figures simply reads the messages, it is obvious which are good ones (i.e., communicate effectively) and which are not. But when one does this, he is simply taking the role of the listener in one of our experiments. What we need is a measure which we can apply to the messages themselves and which does not take into account the choice response of a real or role-playing listener. We know of no measure which does this precisely. But there are some which, it may be argued, are relevant to it.

The work of several sets of investigators (e.g., Brown & Lenneberg, 1954; Krauss & Weinheimer, 1967; Krauss *et al.*, 1968; Lantz & Stefflre, 1964) provides some direct and indirect evidence for the proposition that the communality of names given to stimuli is a good index of how adequately the stimuli can be communicated. Brown and Lenneberg found that differences in the communality of names given to color chips was a good predictor of how accurately the colors could be identified on a delayed recognition task. Lantz and Stefflre showed that the same communality index was correlated with communication accuracy. Krauss and Weinheimer demonstrated that the communality index varied appropriately with the stimulus context in which the referent stimulus was set. And, in the experiment referred to above, Krauss, Vivekananthan, and Weinheimer found that a rough analog of the communality index varied depending on whether a speaker's manipulated encoding condition was public or private. We would not argue that a communality index is the best measure imaginable to be applied to our data. Rather, it is the best one we could think of, and there is at least some rationale for its use.

Therefore, we calculated from our transcripts the proportion of conceptual responses that were not unique—that is, were used by more than one speaker. We did this separately by grade and figure. This proportion will serve as our communality index. The results of this analysis are displayed in Figure 7. For kindergartners the proportion of conceptual responses given by more than one speaker was relatively low, less than one in five. And recall that this is so despite the fact that kindergartners as a group give a relatively small number of different conceptual responses to begin with. Clearly these responses, few in number though they may be, are a remarkably heterogeneous lot.

Now let us look at the data for fifth and eighth graders, bearing in mind the fact that the total number of conceptual responses given by subjects in these two groups is roughly the same. For eighth graders, about 35% of the conceptual responses given are used by more than one speaker; for fifth graders this figure drops to 28%—not a very large difference perhaps, but nearly as large as the difference between fifth graders and kindergartners.

Sometimes qualitative data are more compelling than summary statistics. When eighth graders communicate with kindergartners, 80% of them use the conceptual response "flying saucer" in talking about Form 1. Only 20% of the fifth graders do so. We know that "flying saucer" is a pretty good name for this figure and it is the modal response for adults. Another example, 70% of our eighth graders use the conceptual response "triangle" as part of their message for

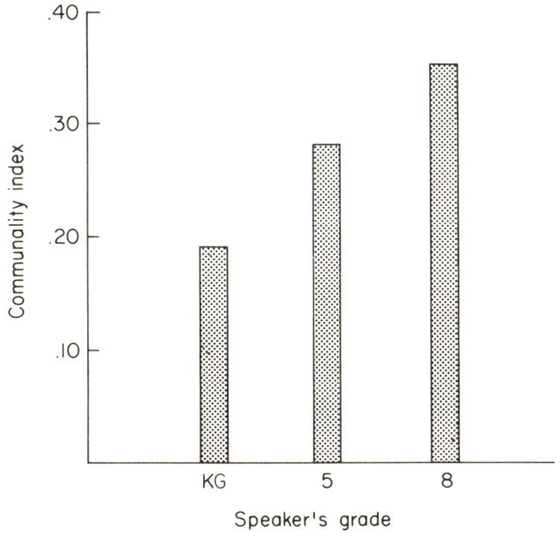


**Fig. 7.** Proportion of conceptual responses used by more than one speaker.

Form 3, contrasted with 20% of the fifth graders. Across all figures the modal frequency is greater in five out of six cases for eighth graders compared to fifth graders.

These data, consistent though they are with our expectations, provide little in the way of direct confirmation for any theoretical position. There is at least one rather tenuous assumption involved and the experiment is not sufficiently well controlled to answer the questions we would like to put to it. What we need, and what we hope experiments presently in progress will provide, is a direct estimate of the repertoires of children of different ages together with some estimate of their ability to discriminate among the alternatives available to them.

The rather meager yield from each of our experiments puts me in mind of the story told by Sholem Aleichem about the legendary town of Chelm. It seems the town elders decided to hire a watchman to stand at the town's gates to announce the arrival of the Messiah. After several months at his job, the watchman complained that his salary was inadequate. "True," replied one of the elders. "But the work is steady." It seems to me that understanding the cognitive processes which underlie the child's developing ability to communicate will provide steady work for psychologists for some time to come.

## REFERENCES


Brown, R. *Social psychology.* New York: Free Press, 1965.

Brown, R. In the beginning was the grammar. *Contemporary Psychology,* 1968, vol. 13, 49-52.

Brown, R., & Lenneberg, E. H. A study in language and communication. *Journal of Abnormal and Social Psychology,* 1954, vol. 49, 454-462.

Cohen, B. D., & Klein, J. F. Referent communication in school age children. *Child Development,* 1968, vol. 39, 597-609.

Flavell, J. H. *The development of role taking and communication skills in children.* New York: Wiley, 1968.

Glucksberg, S., & Krauss, R. M. What do people say after they have learned how to talk? Studies of the development of referential communication. *Merrill-Palmer Quarterly,* 1967, vol. 13, 309-316.

Glucksberg, S., Krauss, R. M., & Weisberg, R. Referential communication in nursery school children: Method and some preliminary findings. *Journal of Experimental Child Psychology,* 1966, vol. 3, 333-342.

Krauss, R. M., & Glucksberg, S. The development of communication: Competence as a function of age. *Child Development,* 1969, vol. 40, 255-266.

Krauss, R. M., Vivekananthan, P. S., & Weinheimer, S. "Inner speech" and "external speech": Characteristics and communication effectiveness of socially and non-socially encoded messages. *Journal of Personality and Social Psychology,* 1968, vol. 9, 295-300.

Krauss, R. M., & Weinheimer, S. Concurrent feedback, confirmation and the encoding of referents in verbal communication. *Journal of Personality and Social Psychology,* 1966, vol. 4, 343-346.

Krauss, R. M., & Weinheimer, S. Effect of referent similarity and communication mode on verbal encoding. *Journal of Verbal Learning and Verbal Behavior,* 1967, vol. 6, 359-363.

Lantz, D. L., & Stefflre, V. Language and cognition revisited. *Journal of Abnormal and Social Psychology,* 1964. vol. 69, 472-481.

Piaget, J. *The language and thought of the child.* New York: Harcourt, Brace, 1926.

Werner, H., & Kaplan, B. *Symbol formation.* New York: Wiley, 1963.

## DISCUSSION

It appears from the material Dr. Krauss has presented that eighth graders are better at communicating with kindergarten age children than fifth graders are, a participant said. Why is this so? What causes a person to modify his description in a way that will reduce uncertainty in his listener? What is the nature of the learning process through which a person develops the ability to regulate his communication in response to the social demands or needs of the listener?

This is the question to which we do not yet know the answer, Dr. Krauss responded. Many factors may affect the way a person communicates. An older child has not only had more experience communicating, but he also has had broader experiences with the world in general. He has available a lot of knowledge that a younger child has not. For example, a kindergartner will repeat a description of something over and over, even if this description does not get his message across, while an older child will modify his description. This may reflect differences in their respective verbal repertoires, but, it also may reflect differences in the communication demands placed upon them in the past. The young child has been exposed mostly to his parents and siblings. In one sense, his own home may be a poor supportive environment for him to learn how to communicate. So much of the meaning of what he says is given in the context that it often is hardly necessary for him to say anything. If you really want to teach someone to communicate, you should not put him in with people who share the same kinds of experience he has and who already have a fair idea what he is going to say. The older child has broader exposure to people with ideas and viewpoints different from his own.

It is not always true that parents understand what their young children mean, a participant pointed out. It is not unusual to find an older sibling or the brighter one of a pair of twins acting as interpreter for a less articulate child. In some cases this mechanism is continued because the "translator" finds it an effective means of controlling people.

Even though the eighth graders are better at communicating with kindergartners than the fifth graders are, neither group of the older children seems to improve much over time in ability to communicate, a participant said. Would it be useful to test the effects of modifying the instructions given the older children? All of us adapt our conversation, our vocabulary, our grammatical structure to accord with our expectations of the listener.

It appears that the eighth graders communicate better than the fifth graders primarily because they are simply better communicators for some of the reasons cited, Dr. Krauss replied. However, the effects of modifying instructions, as suggested, is being tested with adults. We know already that the adult distinguishes between encoding something for himself and encoding it for someone else. What we need to know now is how he uses information about the listener. How do his messages differ when he is talking to an urban ghetto child or a middle class child? His wife? A friend? Also, what does he perceive about this person that makes him think one means of communication will be more effective than another?

The question of what expectations the communicator has of the listener is an important one, a participant said. We know that the expectations teachers have of the children in their classes affect the accomplishments of the children. Rosenthal's studies show that a teacher who thinks a child is bright does different things with him than she does with a child she thinks is stupid, even if both children are actually selected at random. For example, she may use a limited vocabulary with the "stupid" child which does little to enhance his vocabulary.

Some experiments with artificially constructed dyads of black lower class children and white middle class children showed two interesting outcomes, Dr. Krauss said. First, the white middle class children communicated better to the black lower class children than the other black lower class children did. Second, when the white middle class children were the listeners, they understood the messages of the black lower class communicators better than black lower class listeners did. This study should be interpreted with caution, because it would be very easy to jump to wrong conclusions, but it does raise some interesting questions about the glib assumptions people make about white middle class teachers being unable to talk to their lower class black students.

Dr. Krauss' findings seem to be in line with some of Leon Eisenberg's findings, a participant said. Eisenberg asked speakers from four classes—white middle class, black middle class, white lower class, and black lower class to pronounce certain sounds; then listeners in each category were to repeat what they heard. The study has been criticized because connected discourse was not used, and further studies are needed. However, the study did show that the white middle class speakers were understood best by all four groups of listeners, and the black lower lower class speakers were poorly understood even by black lower class listeners. There was not the kind of interaction that people assume exists in which each person understands his own kind of person more easily than he understands other kinds of people.

What does the fact that people modify their descriptions tell us about language as an internal, self-regulatory mechanism as well as a regulatory mechanism in social communication, a participant asked. Does the kind of modification they use give any idea of the way language is organized within the individual?

Probably the best way to test this would be to use a set of stimulus materials over which the investigator had good control, Dr. Krauss replied; that is, figures or objects which immediately suggested one obvious label. Then one could analyze where people went next in their descriptions.

The tendency of kindergartners to describe an object by tracing its shape in the air even though they are visually screened off from the listeners is somewhat reminiscent of a study of urban direction-finding, a participant said. The study involved asking directions of strangers on the street. If their initial answers were not understood, the people who had been asked for directions first started waving their hands, then started tracing a fairly detailed topographic representation.

People also adapt their verbal strategies in trying to give directions, Dr. Krauss responded. A student carrying a concealed tape recorder asked strangers on the streets at Cambridge, Massachusetts for directions. He asked his respondents one of three questions. (1) "Can you tell me how to get to Central Square?" (2) "I am from out of town. Can you tell me how to get to Central Square?" (3) He asked question (1) in a rural Missouri dialect. The recordings were analyzed in terms of the number of words in the response and the number of units of information in the message. It is not surprising that when a questioner mentions that he is from out of town, he receives more information. He is really asking for it; in effect, he is saying, "I don't know very much; you will have to tell me more." But use of the dialect also produced much more information than the same inquiry uttered in a "general American" dialect. People immediately adjusted to the idea that this was a person who would need more information. The amount of information given did not always increase the chances of finding the right route. Sometimes the information given was not very useful; but it did suggest that respondents were making an extra effort to communicate.

Do the studies show that there is a point of diminishing returns in communication between adults and children, a participant asked. Does increasing age lead to better communication in all interactional circumstances, or do you run into a communication lag at some point? Does communication shut down, for example, between a child and adult when the child reaches the "rebel" age group; and if so, is this an interactional effect of the age combination?

It would be difficult to demonstrate this with the stimulus materials used in the studies, Dr. Krauss said. They do not really provide a great deal to talk about, as do topics like drugs or sex. However, this raises a point about another popular misconception. We hear a lot about lack of communication between parents and children or between members of different social groups. As an alternative, one might suggest that in fact such individuals communicate very well; rather they simply fail to agree. Neither group wants to hear what the other is saying, so they reject the appropriateness of the message. But this is quite different from failure to communicate.

Communication is a two-way process involving the listener as well as the communicator, a participant said. Could the failure to improve communication over time in a series of trials reflect poor listening rather than poor communication? Does the good listener modify his behavior more than the poor listener?

The data do show that some people are more reliable and accurate listeners than others, Dr. Krauss said, but it has been difficult to relate such correlations to other factors. The data also suggest that older children are more "active" listeners than younger ones. Younger children tend to accept rather passively what is said to them even when they realize it does not make sense. Often they seem to accept without question the notion of "doing their best" with the faulty and misleading information available to them.

Nathan Maccoby has done some studies of what children's faces reveal about how well they understand something that is said to them, Dr. Eleanor Maccoby reported. A teacher is asked to teach something to a child, and the child's face is photographed. He is then tested to see whether he understood the point. The teacher is shown the movie and asked to "read" the child's face to determine whether he had understood. These results are compared. At first, the teachers score just at chance, but later they improve. This suggests that many people do not automatically know how to read nonverbal clues but that they can learn to do this.

People react to other kinds of nonverbal cues as well, Dr. Krauss said. For example, if a person is talking on the telephone and his listener becomes absolutely silent, the speaker becomes uneasy. He starts asking, "Are you there? Do you understand me? Has the line gone dead?" He also becomes uneasy if a delay is inserted into the telephone line so that the listener's response is transmitted a second or so later than it would normally be. The speaker tends to interpret this as inattentiveness or unresponsiveness on the part of his listener. Often he reacts by providing additional information or becoming redundant. It is difficult to tell whether he does this because he thinks the "slow clod" at the other end of the line needs more information, or because a person tends to keep giving information until the listener shuts him off by saying, "Uh-huh," or "Yes, I understand," or something else which explicitly indicates that he does not need more information.

During his presentation, Dr. Krauss had pointed out that for purposes of his study, it was important that the figures used have a low level of communality; that is, that each figure would elicit a lot of different names from the people who were asked to describe it. The index of communality used was the proportion of conceptual responses given by more than one speaker. Evidence from a number of studies, he said, shows that the communality of names given to stimuli is a good index of how adequately those stimuli can be communicated.

Two participants suggested possible methodological refinements in Dr. Krauss' studies. In response to Dr. Krauss' discussion of the index of communality, a participant suggested that a less crude index might be applicable, using techniques deriving from information theory.

Another participant suggested that it might be useful to scale the responses children make in these experiments with a multidimensional technique to determine similarities. This method might produce clusters of similar items in which common themes might be identified. This could be done separately for each group so that changing strategies could be examined on a qualitative basis. It might suggest the kinds of instructions that could be given the subject and might also suggest additional points that need to be studied. Additionally, it might uncover dimensions that have some psychological cohesiveness. For example, the data might show that kindergartners understand dimensions and kinds of descriptions which they do not use themselves. One could then ask, what is it that communicates to them? What cognitive strategies are involved?

A considerable amount of data of this type is already available, Dr. Krauss said, but some of these ideas are certainly worth considering.

Many of the participants asked for additional information about points Dr. Krauss had mentioned. Several asked questions about the kinds of names subjects assigned to the figures, the degree of communality, and why the subjects appeared to choose certain types of names. At what age is there a clear divergence between descriptions made for self and those made for social purposes, a participant asked. Is there greater communality in the names given for social use than for personal reference?

Are the names people give the figures more conceptual or descriptive? Do they say a figure looks like a specific object such as a milk jug or a shirt, or do they say it is the one with the two oblong pieces sticking up on the sides? Does this vary with age; that is, do children tend to use more pictorial images and adults more symbolic and conceptual images? Do they use whole labels or relationship labels; for example, do they describe number two as a single unit of some kind or as "two hangers"?

Do any of the subjects describe the unique qualities of the figures? For example, there is only one figure that is completely open; only one that is partially open and partially closed; only one that has a threefold symmetry; only one that is both closed and symmetrical; only one that is closed, asymmetrical, and has a straight line.

It is evident that the figures can be described in terms of many kinds of relationships, and some of them are very tricky to grasp. To understand some of the responses of the children, you have to get inside the subjects' minds to know what it was they abstracted. This relates to Piaget's point about language not being very communicative at an early age. Children have a great deal of language knowledge. For example, they apparently have no difficulty with the idea of naming the figures; it seems logical to them that they should have names. But they often describe the figures idiosyncratically; the child says what something looks like to him without regard to the point of view of anyone else.

It has been difficult to determine what influences the kinds of names the subjects give the figures, Dr. Krauss replied, although this does appear to vary with age. Older people virtually always liken a figure to something and then go on to qualify the comparison. They will say, "This figure looks like a milk jug except that . . . ." Young children often describe the figures in terms that are personally meaningful to them but are poor terms for communication purposes. One child, for example, consistently called one of the figures "a sheet," and the other child was never able to identify what he meant. Later, when asked which one was the sheet, the child pointed to figure number three and said, "Haven't you noticed, when you wake up in the morning, the sheet has wrinkles in it, and sometimes the wrinkles look like that?" This is really a rather poetic interpretation, but not good for communication unless the listener knows the basis of the description.

This system of idiosyncratic labeling may not be particularly good for communication purposes, but it may be very useful in other types of intellectual performance, a participant suggested. In a recent experiment, Alvin Duress has been attempting to discover what conditions make it possible for a child to copy the action of another person. He has the subjects look at what another person is doing. He asks the children in one group to describe the action in sequence (the hands start at the level of the shoulders, they move together toward the front of the body, etc.) He asks the other group to give any kind of summary or "encapsulating label" for the action (a bird lighting on a pond). After a lapse of time, he asks the subjects to reproduce the motion. Sometimes the action has several steps in sequence so that it is possible to tell not only whether the subject has recalled all the components but whether they are in the right order. The results show that the encapsulating label is much more efficient than the sequential description both in relation to the total number of components recalled and to getting them in the right order.

# REGULATORY FUNCTIONS IN COMPUTER MODELS

*John C. Loehlin*
The University of Texas at Austin

The topic of this chapter is self-regulatory functions in computer programs. Since this is supposed to be a volume on *human* self-regulation, perhaps a word of preliminary explanation is in order.

The basic argument is this: Computer programs (in computers) can exhibit complex and interesting behaviors, resembling in many cases those emitted by humans. In particular, computer programs can display a variety of complex symbolic behaviors. Furthermore, computer programs are (relatively) convenient to study, since the details of their organization are accessible in a way that the details of the organization of the information and control systems governing human behavior are not. Now it is quite possible, of course, that the self-regulatory functions inherent in computer programs may not closely resemble those controlling similar behaviors in people, in which case we are somewhat in the position of the well-known drunk looking for his watch under the streetlamp because the light is better than in the place where he lost it. We may console ourselves, however, with the thought that any enrichment of our stock of ideas about self-regulatory processes may be worth having; after all, the drunk by searching in a good light is enhancing his chance of finding *something,* and if he goes home with a wallet or a pair of cufflinks instead of his watch his effort is not a total loss. Also, some of the computer programs I will be discussing were deliberately designed to mimic certain aspects of human behavior, and therefore embody implicit theoretical notions concerning human self-regulatory processes.

## COMPUTER PROGRAM STRUCTURES

We will begin with a very brief review of the basic ingredients of computer programs—the raw stuff of which the systems we will be discussing are built.

A computer can be viewed as a huge set of pigeonholes, the *computer memory,* in which appropriately coded data or instructions to the computer can be *very* rapidly stored and retrieved. Operating with this memory is a *processor* containing electronic circuits capable of fetching numbers from the pigeonholes and carrying out simple operations with them—adding, subtracting, or comparing them, for instance. A computer system will also contain some *input–output devices*—card readers, tape drives, printers, and so on—for getting information into or out of the system. A *computer program* is merely a sequence of coded instructions, themselves stored in the memory, each of which tells the processor to carry out some one of the operations in its repertoire.

The simplest form of program is a linear string of instructions which the processor executes one after another from beginning to end. The program is started by giving the processor the location of the first instruction and turning it loose. The successive instructions are often located in contiguous pigeonholes in memory, but this is not essential—in some systems instructions may be widely scattered about, with each containing an explicit indication of where to go to pick up the next instruction. In either case, the end of the program is marked by a stop instruction which releases the processor. Thus we have here the simplest form of regulation: a start, sequencing, a stop.

If this were all there were to it, the computer might still be a useful and practically important device, because of its ability to carry out long strings of instructions at incredible speeds, but the computer program would be of limited interest to us in discussing models of self-regulation. But this isn't all there is to it. One additional simple innovation makes possible virtually unlimited additional complexity in computer programs. This is the *conditional branch,* or decision instruction. This is a choice point in a series of instructions, where the processor may go one way or the other, depending on some condition that exists at the moment of decision; e.g., the branch instruction may specify that if the value in a certain location is zero, the processor should go to one place for its next instruction; if it is non-zero, it should go somewhere else. Since the program may itself have earlier set the critical number on the basis of prior computations, it is obvious that the programmer when writing the program doesn't know how it will actually proceed in a given instance: he simply assigns to the program the authority to make a local decision on the basis of specified criteria when the moment arises.

The conditional branch immediately opens up virtually unlimited program complexity, since several simple decisions can be sequenced to yield a complex decision, and any number of such decisions can be linked together by linear program segments to yield a network of arbitrary complexity. Since the conditions governing any particular decision can be made to depend on the joint outcome of numerous previous actions and decisions, and since these decisions may include decisions to read in and react to information from outside the system, it is clear that extremely varied and subtle effects can be achieved. As Marvin Minsky (1968) puts it, while one can correctly think of a computer as a device that executes a series of simple instructions, "it is just as correct, and much healthier, to think of the computer as a nutrient medium—a garden—for maintaining unthinkably tangled webs of structure [p. 3]."

Even in gardens one can detect simpler structures, and this holds true for the tangled growths of computer programs as well. Let me mention four such structural features: the iterative loop, the subroutine, recursion, and the interrupt.

In the loop, a sequence of instructions loops back around and reenters itself. A decision point must be provided somewhere in the loop to provide an exit from it—otherwise the processor will simply proceed serenely around and around the loop until the allotted computer time runs out. One may provide a decision point that exits when the loop has been executed a fixed number of times. For example, the loop may include instructions that add one to a number on every pass around, and each time the decision instruction compares this number with the desired number of cycles to see if it is time to quit yet. A more interesting regulatory device, from the standpoint of our present concerns, is a decision based on whatever it is the loop is supposed to accomplish. Here the decision is based on the discrepancy between what is sought and what has so far been achieved, and the program steps are designed to diminish that discrepancy on each cycle. When the result is "good enough," according to a specified criterion, the program exits from the loop at the decision point. Many numerical programs in computers operate in this way, by obtaining successively better and better approximations to some desired value, and quitting when the value has been pinned down to within a desired range of accuracy. As a psychological model, the iterative loop is the basis of Miller, Galanter, and Pribram's (1960) well-known TOTE (*T*est-*O*perate-*T*est-*E*xit) unit. To take their example of hammering a nail: the loop includes the elementary instructions necessary to strike a blow with the hammer, and the exit decision is based on whether the nailhead is down to the wood yet; thus the loop is repeated until it is. For real-life hammering, of course, one would want to provide several possible exits from such a loop, to provide for termination in such cases as bent nails, broken hammers, and noticing that the nail has penetrated the wood and entered the surface of the dining-room table.

As Miller, Galanter, and Pribram note, such loops may be arranged in hierarchical fashion: a major loop may contain one or more subloops, each of which may in turn contain further subloops, all of which must be satisfied (working from the inside outwards) in order to exit from the main loop. Thus the nail-hammering loop may include little subloops regulating the performance of the individual hammer stroke, and the nail-hammering may itself act as a subloop in a broader bird-house building program.

Another ubiquitous structural feature of computer programs is the *subroutine.* This is an independent subportion of the program that can be entered from various points of the main program, with return to the appropriate point of the program upon completion of the subroutine. Normally, a subroutine carries out a specialized and carefully specified operation. Its relative isolation from the rest of the program means that it can be constructed and tested as a separate unit, and can serve as a component in many different programs. Indeed the program writer can often use subroutines written by others without ever knowing *how* they accomplish whatever it is they do, so long as he knows exactly *what* they do. Any large computer system has dozens of standard subroutines available—to take square roots, generate random numbers, handle input and output operations, and the like, and a user might go for years without ever having occasion to look inside one of them.

A multilevel hierarchical organization is common with subroutines as well as with loops. A main program may call various subroutines, which in turn call lower level subroutines, and so on. In fact, most large programs are highly subdivided in this way—the main executive program may be little but a series of calls of major subroutines representing broad subdivisions of the program. These major subroutines may themselves essentially be schedules for the calling of a second level of subroutines which carry out the actual details of the operation, and these in turn may call on specialist subroutines from time to time for highly stereotyped operations.

Suppose that a subroutine hierarchy loops back upon itself, that is, subroutine A calls subroutine B, which calls subroutine C, which in turn calls subroutine A from within itself, creating a potential infinite regress, as A calls B calls C calls A again, and so on. This is called *recursion,* and it can be useful, provided some exit from the sequence is built in. Thus a grammatical analysis program, finding a sentence embedded within the sentence it is working on, might call the main program as a subroutine to analyze the embedded sentence, on successful completion of which it returns to its place in the original sentence and proceeds on its way. If during the analysis of the embedded sentence a sentence embedded within *it* were encountered, the process could be repeated, to any desired depth. The only trick in such operations is to keep track of enough information about the status of things at each level of the recursion so that processing can be resumed properly when work is finished at the lower level. This introduces an appreciable overhead cost in memory space and computation for each additional step of recursion, so that in practice the process is self-limiting. Such constraints are probably even more severe when brains do the computing, because of the sharp limits on available immediate memory.

One additional control principle may be mentioned: the *interrupt.* This is a much-used process in computer systems, and it may be used within a program as well. The interrupt is a provision for allowing a higher priority operation to interrupt an ongoing process of lower priority. Thus in a computer-controlled experiment it may be necessary to interrupt momentarily the ongoing computing in order to store an incoming signal from the subject's response key, at an unpredictable point in time after a stimulus is presented to him. One way of achieving this is to have the program check after every few instructions to see if there is an input signal coming in; if not, it continues computing; if there is, it branches to a subroutine that records the input and carries out any immediately necessary processing before returning to the main program.

Herbert A. Simon has given the concept of the interrupt a key role in his treatment of motivation and emotion (Simon, 1967). An unsatisfied need, growing in intensity, may lead to an interrupt of an ongoing action sequence and its replacement by a program of action relevant to the critical need. A "loud" external stimulus—for example, a sudden noise—can act in essentially the same way. Similarly, an emotional association may interrupt an ongoing process of thought or interpersonal behavior, sending the process temporarily or permanently off in a new direction.

## EXAMPLES OF SELF-REGULATION

With these simple forms of program structure before us—the linear sequence, the conditional branch, the loop, the subroutine, recursion, the interrupt—let us turn to a consideration of some examples of self-regulatory processes as they operate in computer programs whose behavior is complex and humanoid enough to be psychologically interesting or which have deliberately been designed as simulations of human psychological processes. Included will be a simple "personality" model called Aldous; a more complex model of a neurotic personality constructed by the psychoanalyst Kenneth Mark Colby; an M.I.T. chess-playing program (Greenblatt's); and a robot which trundles around at the Stanford Research Institute. Finally, a program will be described which does very simple sensory-motor learning in something of a Piagetian fashion. In each instance enough will be said about the program to suggest something of its general mode of operation, and then we will look more closely at self-regulatory processes within it.

*Aldous*

This computer program was intended to model some simple aspects of personality (Loehlin, 1962, 1963, 1968a). The program, named Aldous, is capable of developing attitudes toward a moderate number of objects or situations (say, 10 to 1000). Each "attitude" is represented by a storage location in the computer which contains numbers representing the degree to which Aldous is predisposed to experience each of three emotions (love, fear, and anger) toward the object, and an indication of the amount of experience on which this attitude is based.

Interacting with this set of attitudes are a number of subroutines; a recognition routine that locates relevant attitudes in response to a stimulus input; an emotional reaction routine that generates an emotional response which takes into account both relevant attitudes and the model's current emotional state; an action routine that translates the emotional response into an appropriate action; and a learning routine that modifies the attitudes in memory in the light of the consequences of the action. Additionally, Aldous contains an introspection routine that can answer a small set of appropriate queries about its experience (e.g., "How did you feel about that last situation?") with a verbal report which is capable of communicating somewhat more detail about what is going on. One can, for example, infer conflict from the model's behavior, but the introspective report will indicate which emotions are involved in the conflict.

External to Aldous are other programs in the computer which represent the external world: These programs input stimuli to Aldous, accept outputs from him, determine what consequences should ensue from Aldous's actions, and so forth. Experiments with Aldous may be preprogrammed, i.e., based on an input tape of a specified sequence of situations, or may develop dynamically. An interesting version of the latter involves two copies of the model in the computer arranged so that the actions of each in turn serve as the stimuli input to the other. Various experimental possibilities exist. For example, one may start the two models off with different attitudes, and then see what happens as they interact. In such experiments, often both models will wind up with attitudes quite unlike the ones with which either started out. A friendly and a fearful model might end up in mutual hostility, for instance, as a result of the frustrations produced in their interaction (Loehlin, 1965). Another sort of experiment involves varying certain parameters of the models and observing the effects on the models' interaction. Such parameters could include, for example, rate of learning, emotional lability, accuracy of recognition, or the relative weight given to enduring attitudes and current mood in generating emotional responses. They might be considered analogous to personality traits or stylistic traits of the models. Such changes of parameters have effects which are often not easy to predict, although intelligible in retrospect. Often the surprising thing is how little difference large changes in parameters seem to make to the overall behavior of the system. To look at this the other way round: although this has not actually been tried, it probably would be extremely difficult to infer accurately the values of Aldous's internal parameters from observing moderate amounts of his behavior (or verbal reports). This for a rather simple system whose structure is known. How much more difficult, then, to do what psychologists often purport to do: make the same sorts of inferences for much more complicated systems whose structure is poorly understood—namely, human persons.

But back to our central theme. In what sense does this model contain self-regulatory mechanisms? Are they built-in or acquired via experience? What role do they play in Aldous's developmental history?

Self-regulation in Aldous is clearly very primitive, but there is a general tendency toward stability of his behavior over time. There are two main stabilizing mechanisms. First, attitude change in Aldous in an inverse function of amount of experience, so that development of his attitudes tends to produce increasingly stable responses toward particular environmental events. And second, the persistence of his emotions from trial to trial tends to smooth out his responses over time in an erratic environment.

A third regulatory mechanism may be mentioned, which tends to simplify the model's emotional response to events. There is a degree of mutual incompatibility of emotions built into the emotional reaction subroutine, so that the dominant presence of any one emotion in a situation tends to diminish the strengths of the others.

All three of these regulatory devices are "innate," i.e., programmed into the system. However, their role in governing the behavior of the model is in various degrees dependent on the model's past experience, both recent and remote.

But on the whole, one must say that Aldous displays a minimum of self-regulatory mechanism. He does not contain a separate homeostatically based system of primary drives, his perceptual system does not achieve object constancies, and his effector system is not coordinated to a physical environment by a moment-to-moment interplay of plans and feedbacks.

Aldous, however, is a very primitive program, originally written in machine language for a small 4000-word computer; indeed at the time he was first programmed his author had never actually run a program on a computer! Let us move on to consider more complex cases.

### *Colby's Model of a Neurotic Process*

A considerably more elaborate program designed to represent personality processes, in this case neurotic ones, is the model programmed by Kenneth Colby at Stanford (Colby, 1963, 1965; Colby & Gilbert, 1964). In addition to representing objects (in this case person objects like father, mother, self) the model represents beliefs about relationships among these objects, beliefs such as "I love mother," "Father despises mother," "Mother attracts older men," and the like. Associated with each of these beliefs are several numbers representing such properties as the current momentary degree of emotional investment in the belief (its "charge"), the belief's habitual or enduring emotional importance, the degree of certainty with which the belief is held to be true, and its supportive relationships with other beliefs. (One reason for believing that "Father despises women" may be the belief that "Father despises mother"; weakening the credibility of the latter belief will tend to weaken the credibility of the former as well.)

Much of the active processing in Colby's model represents the conflicts that occur among strongly charged beliefs. Thus if an attempt is made to express the belief "I hate father" when the belief "I ought to love father" is present and active, a conflict exists, and various transforming subroutines, the analogues of psychoanalytic defense mechanisms, may be called upon to deal with it. These subroutines carry out various distorting operations on beliefs. For example, "I hate father" might be changed to "Father hates me" in a projection-like transformation. The successfully transformed belief is expressed, and also added to the belief structure, which means that it may in turn generate new conflicts later on.

The model also contains a number of indicators called "monitors"—a danger monitor, an excitation monitor, an esteem monitor, etc.—which reflect the character of the ongoing processing in the system. These feed into an overall "well-being" monitor. Successful processing of beliefs tends to decrease the level of the danger monitor and raise the level of the well-being monitor, but if efforts at transformation fail to yield expressible beliefs the danger level rises and the sense of well-being drops. The other monitors are sensitive to various aspects of the processing. For instance, the esteem monitor keeps track of the system's self-esteem—chiefly governed by the fate of beliefs (such as "I ought to love father") which express felt obligations of the self.

In addition to expressing the emotional state of the model, the monitors carry out regulatory functions. For example, if the well-being monitor falls below a certain threshold level, the current topic of processing is dropped, and the model's attention shifts to a new (and hopefully less conflict-ridden) area of its attitude structure, with a concomitant rise in the sense of well-being. Thus the monitors may act as interrupt devices.

The degree of conflict resulting from contradictory beliefs is a function of the current charges on the beliefs and the conviction with which the beliefs are held. The current charge is sharply reduced by the expression of a belief, and reduced to a lesser extent by the expression of a transformed version of the belief. The more distortion required to express a belief, the less effective its expression is in reducing the charge on the original belief. Current charges on all of the beliefs in the system build up by a fixed fraction of their long-term values every time the topic of processing shifts in response to monitor interrupt.

Thus, in terms of self-regulation, we see in Colby's system a series of positive and negative feedbacks, some of continuous and some of threshold character, which appear overall to yield a somewhat unstable system; that is, a system which in the absence of external input should tend to go eventually to either very low or very high levels of internal stress. For if conflict levels are tolerably low, the system should be able to express most of its beliefs with relatively little distortion, leading to increasingly favorable monitor levels and increasingly low charges on beliefs, in turn facilitating the expression of further beliefs. On the other hand, if conflict levels become relatively high, greater distortion will be required to express beliefs, leading to more frequent failures and a worsening of monitor conditions, and the beliefs that *are* expressed in distorted form are relatively ineffective in lowering current charges. While the shift of topic provides a protective mechanism for avoiding immediately dangerous monitor levels, frequent shifts will tend in the long run to raise charges throughout the attitude system. Also, the addition of distorted beliefs to the system may make things increasingly worse by introducing further potential conflicts. However, it would seem possible, at least in principle, that the model might sometimes wind up with a set of mutually consistent (though distorted) beliefs, a stable delusional system, if you will, which could also lead eventually to low charges and favorable monitor levels.

Provisions for long term "therapeutic" change in the model involve inputs by a "therapist" which direct the model's attention to certain topics, and a provision for gradually weakening the credibility of not-too-highly charged conflictual beliefs if they lack support elsewhere in the attitude structure. Presumably such "therapy" would best proceed by first directing the model's attention to very mildly threatening topics which it could readily handle, and as internal conditions in the model improved, leading it gradually on to confront increasingly more severe conflicts—all the while with an eye to weakening the credibility of conflictual beliefs, beginning with relatively peripheral ones and gradually chipping away the support of the attitudes central to the model's conflicts.

Most of the regulatory structure is programmed into the model to a greater or lesser degree; however, the use of particular transformational mechanisms to handle conflicting beliefs is subject to learning. In general, the transformational mechanisms are arranged in order of the amount of distortion they produce, and more serious conflicts call forth more severe distortions. However, the successful use of a particular mechanism increases the proportion of the danger range allotted to it, so that a characteristic style of defense may be developed by a particular model on the basis of its experience. Since many factors enter into the determination of each success or failure of processing, such trends are not easily predictable in advance.

*Greenblatt's Chessplaying Program*

The two computer programs discussed so far were designed to model or simulate human behavior, especially those attitudes and emotional traits which we tend to classify under the amorphous label "personality." Consider now a rather different sort of program, designed to carry out a particular intellectual task, in this case, playing chess, without deliberate effort to mimic the manner in which humans do it. As it turns out, the way in which the program plays chess bears several resemblances to the way people seem to play chess, although there are some differences, too, as we shall see.

The chessplaying program was constructed at M.I.T. by Richard D. Greenblatt and various collaborators (Greenblatt, Eastlake, & Crocker, 1967). It is a decent amateur player, winning about 80% of its casual games. It has played in several local and state tournaments, and while it has not won any championships, it has won some tournament games and has a modest but respectable chess rating. Under the name MacHack VI it is an honorary member of the U.S. Chess Federation and the Massachusetts Chess Association. It plays under regular tournament rules and time limits, with the assistance of a human operator who types the opponent's moves into a teletype and makes the moves on the board that the computer types back.

Now there is no particular difficulty in programming a computer to play legal chess. Nor is it much of a trick to write a program that will take a suitable representation of a chess position and find all the legal moves that can be made from it. Hence 20 years ago, almost as soon as there were computers, there were proposals for constructing chessplaying programs to run in them (Shannon, 1950). The basic idea was simple: Consider all legal moves from the present position, one at a time; for each of these, consider all the opponent's replies; for each of these, consider all possible own next moves, and so on, until a sequence of moves is found that leads to a decisive advantage regardless of the opponent's play. All this can easily be programmed in a series of nested loops or recursive subroutines. The only difficulty with this scheme is that since there are on the average around 32 legal moves from a typical chess position (deGroot, 1965, p. 22), the total number of positions to be considered mounts up astronomically if one looks more than a very few moves ahead. (The total number of lines of play is on the order of $32^n$, where $n$ is the depth of the lookahead sequence.) According to the studies of the Dutch psychologist and chessplayer Adriaan deGroot, even fairly weak players will on occasion consider move sequences 8 or 10 moves ahead from a given position (deGroot, 1965, Appendix II). (Incidentally, deGroot found that chess grandmasters do not look further ahead than weaker players, they just look ahead in more profitable directions.) If one were to consider move possibilities exhaustively to a depth of 8 moves (4 own moves and 4 replies), it would require the examination of $32^8$ or about 1.1 trillion move sequences. This is a large number. For example, if your computer is fast enough to consider 200 8-move sequences per second, it would take you something like 175 years of computer time to decide each move. To put it slightly differently, if your program had started playing at the time chess was invented in the sixth century, it would still be in the opening phase of its first game.

To make a practicable chess program, then, it is necessary either to restrict sharply the depth of the lookahead, or to consider only some of the legal moves at each step. Restricting lookahead to only a few moves seems to lead inevitably to glaring blunders in play. On the other hand, selecting only a few moves to consider is compatible with good play *provided* the moves retained for consideration at each step include the strongest moves for each side.

Greenblatt's chessplayer takes the route of selectivity. Built into the program are some 50-odd chess rules-of-thumb about such matters as control of the center, keeping open lines for ones own pieces and blocking the opponent's, avoiding pawn weaknesses, attacking the opponent's weak points, detecting common traps and pins, and so forth. All legal moves from a given position are examined and ranked in order of merit, based on the above considerations. Only the most promising (from the point of view of the side to move) are taken further in the lookahead sequence. The "principal variation," the sequence of apparent best moves for self and opponent, is followed out first to some predetermined depth (if unsettled conditions prevail at this point, such as pieces under attack, the variation is followed further until the situation is clarified). The final position is then evaluated, based on any material advantage to one side or the other, plus a few simple positional considerations. Other alternative moves are then considered for the opponent and the self, working backwards along the principal variation, and abandoning the search along each alternative branch as soon as it becomes apparent that it is either less desirable for the self then the principal variation, or less desirable for the opponent, when he has the move. Since the presumptively strongest moves for both sides are considered first, the examination of other possibilities can often be curtailed sharply as it becomes apparent that they lead to such inferior results for one side or the other that they will be avoided in actual play. Greenblatt estimates that this last procedure cuts down the amount of search by a factor of at least a hundred.

In casual play, up to six alternative moves are considered from any position, although this limit is waived in cases involving checks or threat of checkmate, to allow consideration of all legal moves if necessary. For tournament play, a somewhat larger number of alternatives are considered, 15 at the first level (for self and opponent), 9 at the second, and 7 thereafter.

Greenblatt's program still considers many more possibilities in making a move than human players do, perhaps by a factor of hundreds, but clearly it examines only a minute fraction of all possible move sequences. As its performance indicates, its somewhat more extensive exploration roughly offsets its rather crude evaluation of positions, yielding a creditable level of play by human standards.

While Greenblatt's program was designed for a particular intellectual task, playing chess, there is reason to believe that the mental operations of chessplaying are not entirely dissimilar to those involved in many other forms of problem-solving, in which actions have to be selected and sequenced to achieve some desired result (Newell, Shaw, & Simon, 1958). The problem of seeking for a workable plan in a vast maze of possible plans is common to many intellectual endeavors. The essence of problem-solving skill in such cases is to spend as little time as possible in extensive exploration of blind alleys, and thus have the maximum time available to explore in depth the most promising avenues of approach.

What then are the principal regulatory mechanisms in Greenblatt's program and how were they developed?

Clearly, the key feature is the assessment of the promise of moves, which channels the lookahead process. This is carried out by a routine which evaluates each move in terms of a fairly large number of separate criteria, and combines the values additively with fixed weights. These values are calculated serially, but functionally they operate in parallel, each consisting of an independent evaluation of the move in terms of some property: center control, gain of material, or the like, with the results appropriately weighted and added together to yield an overall assessment of the value of the move. The weights used are based on traditional chess lore, and can be tinkered with by the programmers to improve the program's play. Some other game-playing programs, notably the excellent checker-playing program of Arthur Samuel, adjust their own evaluative weights automatically, on the basis of association with success and failure in play (Samuel, 1959, 1967). Other refinements in Samuel's sophisticated program are the use of different sets of weights for different stages of the game (some properties of positions are evaluated more highly in the opening than in the end game, for example), and the use of nonlinear combinations of properties. However, as Greenblatt's program demonstrates (and for that matter, earlier versions of Samuel's), a simple linear sum of properties can yield quite an effective evaluation *provided they are the right properties,* i.e., that they do in fact reflect the things that really matter in the problem domain.

Besides the evaluation of moves, and the prescribed features of the search scheme, the main factor regulating the lookahead of Greenblatt's program is responsiveness to various features of the position as they develop. Thus the search will go deeper rather than try to evaluate an unstable position with several pieces under attack, or it will go wider to insure that all potential replies to a checkmate threat are considered. The basic program machinery in all these decisions is of course the conditional branch. Have six moves from this position been considered yet? If no, return to evaluate next move. If yes, is there a threat of mate? If yes, continue evaluating more moves. If no, go on to next stage. These decisions are written into the program by the programmer. However, the program has been designed so as to permit easy modification of them, and in fact one important reason for its present level of competence is that the program can be asked to replay a game it lost, with explanations as to why it chose particular moves, and the decision network or the evaluative weights suitably adjusted to take care of the oversight or misjudgment that got it in trouble. Since the program has by now played some hundreds of games of chess, and has undergone a large number of minor revisions and adjustments as a result, it can be said to have learned quite a lot about chess—in a way not entirely dissimilar to a human chess beginner learning by having an expert player go over his games with him. Samuel's checker-playing program is even more autonomous in this regard: It plays through published games of experts and automatically learns to revise its judgments of positions to bring them more into line with the decisions the checker masters made.

*The SRI Robot*

A common feature of the computer programs discussed so far is that they operate in abstract or highly formalized environments. The next program is one designed to act as the brains of a robot operating in a real physical environment. This robot is currently under development at the Stanford Research Institute at the hands of Nils Nilsson, Bert Raphael, and various associates (Nilsson & Raphael, 1967; Green, Wahlstrom, & Forsen, 1967; Rosen, 1968). It consists of a wheeled platform carrying a box of electronic gear, surmounted by a TV camera and optical rangefinder which can be pivoted and swung to scan the robot's environment. The whole works stands about 6 feet high and with a little imagination can be seen as having a faintly hominoid appearance. The robot is currently connected by an overhead cable to its controlling computer, but the cable will eventually be replaced by a radio link. At present, the robot can move freely about a limited environment, a 20 × 30 foot empty room which has some large wooden objects of simple geometric shapes scattered about in it. The initial tasks given to the robot are to move around from one place to another in this simple environment, planning its routes to avoid obstacles. Later, slightly more complex tasks are envisioned, such as "find a wastebasket, and move it to a place from which no doorway is visible." Currently the robot moves things by pushing them.

Many challenging problems in self-regulation are presented by the design of a device of this sort. Not least is the protection of the robot (and the researchers) from the consequences of untoward actions of the machine. In the present case, this has been done by supplying touch reflexes: tactile sensors of the "cat's whisker" type are attached around the robot's base platform, and connected via on-board electronic logic to the drive motors and brakes. If these sensors touch something, the robot stops immediately. It is possible to override the peripheral reflexes by explicit command from the central computer brain, to permit such activities as pushing objects, but, first the robot stops, then it thinks.

Navigation around the robot's environment is planned and controlled with the aid of an internal representation, a kind of map, of the known features of the environment, which is built up and maintained in the central computer memory. This internal map is updated as the robot encounters obstacles or other new environmental features in its travels. Interesting strategic and tactical problems are encountered in planning movements with the aid of such a map. If the apparent shortest path to the goal lies partly across unknown territory, should the robot gamble on finding a clear path, or should it take a somewhat longer route which is known to be passable? How detailed should the initial plan be? If it is not worked out in detail, avoidable difficulties may not be detected. On the other hand, much effort will be wasted if an elaborate plan based on a limited knowledge of the environment is nullified almost at the start by discovery of an unforeseen environmental obstacle.

Obviously, sensible planning decisions in such cases depend on some knowledge of probabilities (how frequently have unexpected obstacles been discovered recently?) and of utilities (what are the relative costs of planning time and travel time?). At present, strategy choices are made according to simple priority rules supplied by the programmer, but it is anticipated that eventually the robot's own accumulated experience will be allowed to play a role in such decisions.

The organization of planning programs in the robot is hierarchical. At the top level a broad decision is made ("proceed on a step-by-step basis"). At the next level this is given more concrete form ("proceed to point *X* and look around to find a suitable point *Y* to go to next"). And then at the lowest level of the hierarchy commands are given to the motors governing the drive wheels and the visual scanning mechanism. The execution of the plans is always subject to automatic interruption from the tactile reflexes, and other interrupts can be written into the control programs—for example, if the visual scan discovers major unexpected features of the environment, it might be desirable to stop and reconsider the route before going further.

In practice, various aspects of the robot's design are proceeding somewhat independently. A basic mobile robot exists, and can carry out simple movements through its environment. Programs have been written to accept task assignments and make reports to the operator in simple English sentences via teletype. Elementary planning capacities have been developed, and subroutines written to accept information from the sensors and incorporate it in a schematic model of the environment. Much effort has been devoted to the problem of recognizing objects in the noisy visual input from the TV camera. Several schemes have been worked on, but so far the robot is still far from being able to recognize dependably even the simple three-dimensional geometric solids which currently serve as the objects in its environment. At present, preprocessing hardware simplifies the visual image somewhat, and a series of programs attempts to select out continuous lines, filling in gaps as necessary; to interpret these lines as the edges of surfaces in three dimensions; and to match the surfaces to the stored images of objects. In theory, such complications as the partial obscuring of objects by one another, and the presence of extraneous lines produced by shadows, floor scratches, and faults in the TV picture can be dealt with, but in practice effective visual pattern recognition has turned out to be among the most refractory parts of the robot enterprise. The everyday human performance of perceptual object constancy is not easily duplicated by machine, even though this would not seem to present any inherent impossibilities.

As indicated earlier, the robot's learning is, so far, largely limited to its acquisition of a model of its environment, although eventually it should be able to modify its planning strategies as well. Its "purpose" at any given moment may be thought of as the task assigned by the operator, or better, as a hierarchy of subgoals descended from this task. Even this oversimplifies the situation somewhat, since a concurrent self-preservation motivation is implicit in the protective reflex system. In addition, the robot has a basic exploratory mode to which it reverts when no explicit tasks are assigned, a mode in which it simply investigates its environment and builds up its internal map. Thus it could be said to have a built-in curiosity motive as well.

*Plummer's Sensory-Motor Learning Program*

The programs discussed so far, while they exhibit a variety of self-regulatory mechanisms, have not for the most part been greatly concerned with the origins of such mechanisms—though in some cases attention is paid to their elaboration and development. I would like, therefore, to say a little bit about a program which attempts to represent some very simple sensory-motor learning. This is a program written by Robert P. Plummer in connection with a Computer Science Ph.D. dissertation recently done at the University of Texas. Plummer was technically working under my supervision, but in fact proceeded largely on his own in this work. The program attempts to do simple sensory-motor learning at the level, say, of a human infant just beginning to manipulate his environment. Plummer has in general been guided by Piaget's description and analysis of such behavior, but does not consider his program as a serious attempt to model Piagetian theory.

Plummer's program starts with a representation of an environment (two-dimensional, for convenience) containing an arm, a visual field, and several objects. The program is presumed to have two senses, vision and touch, and is capable of random movements of its visual field and its arm. Learning in the model consists in the elaboration of a network of perceptual tests and associated actions, somewhat in the manner of the discrimination net in Feigenbaum's EPAM model (*E*lementary *P*erceiver *A*nd *M*emorizer; Feigenbaum, 1961), but with the difference that actions may be interpolated within the network and are not just found at the terminal nodes. Plummer is currently starting his model off with a discrimination net containing three tests. The top test is an "attention" mechanism which determines which sense, vision or touch, is currently receiving the stronger input, and branches to the appropriate test at the next level. These consist of a basic visual test, which ascertains whether the center of the visual field is currently being stimulated, and a "touch reflex," which responds with an arm movement to contact of the side of the arm with an object. The program is capable of constructing new tests (based on randomly abstracted portions of the visual or tactual input), adding them to the network, and associating actions with them. By trial and error the program can come to identify situations from which certain actions will allow a basic test to be satisfied. This results in TOTE-like chains of tests and actions—e.g., visual input in left peripheral vision, move visual field left $x$ units, satisfy basic visual test. New tests are tolerant at first in terms of the matches required to satisfy them, and actions are specified only probabilistically within ranges. With experience, both tests and movements become more narrowly defined. If a reasonably effective procedure fails to develop over a period of time, the test is dropped and a new start made. Whenever the model gets to a blank point in the net, a random movement may be made; if this produces a "striking" perceptual consequence (a marked increase in input), a test based on the consequence is added to the net. Tests so established may themselves act as targets for further learning. Again, if they are integrated into successful action sequences in a reasonable period of time, they are retained, otherwise they are dropped.

At its present stage of development, Plummer's program has successfully (if somewhat laboriously) acquired some visual centering movements, as well as some sequences such as "see arm and object, move arm, touch object." Note that it is capable of acquiring coordinations between vision and touch, even though it has none initially.

I should emphasize that this work is still very much in progress, and it is too early to say much about the capabilities and limitations of Plummer's program. I suspect that it would take it an awfully long time to learn to play chess, or pingpong. But then it takes most human infants quite a few years to learn to play chess, or pingpong. More seriously, the human infant is working with much more complicated biological input and output systems, and this alone must surely guarantee that Plummer's program constitutes a gross oversimplification of early human sensory-motor learning. Still, the basic notion of a system that can start with a very limited decision structure and expand it by incorporating into it aspects of the "causal texture of the environment" is an attractive one.

## FINAL COMMENTS

I have reviewed in this chapter several more-or-less complicated computer programs, and I hope I have at least convinced you that these are interesting objects in their own right. I have argued elsewhere (Loehlin, 1968b), at least three-quarters seriously, that all (or nearly all) running computer programs can properly be considered to be both cognitive and motivational systems. Cognitive, in that they build and revise symbolic representations of events external to the program. Motivational, in that there is an internal decision structure that causes the program to work toward certain ends. Now both the cognitive and motivational properties of a computer program are normally supplied by a programmer, who decrees that the number in memory location $x$ is to stand for the square root of the number in memory location $y$, and that the program will modify $x$ until it is within ±0.0001 of $\sqrt{y}$. However, the meanings and goals are now inherent in the structure of the program, and do not depend on the continued existence of the programmer. In a very real sense, we may describe them as the *program's* meanings and goals.

As we have seen in the various programs we have discussed, quite elaborate systems of meanings can be built up in computer programs, and the behavior of such programs can reflect quite complex systems of values. Indeed, the very flexibility of these programs should serve both as a promise and a warning. A promise, in that our capacity for building models of complex dynamic structures and processes is now extremely great. And a warning, in that this very facility may leave us with models of so many degrees of freedom that we will hardly know where even to begin checking them against data.

As for self-regulatory mechanisms, we have seen a number in operation in the different programs: mostly built-in by the programmer, but sometimes modified and elaborated by the program's experience. The programmers have put these mechanisms in to do what needed doing for the program to behave in the way that it does. How useful some of these notions may be as aids in your own thinking about self-regulatory mechanisms, I cannot, of course, predict. But I hope you have found it at least provocative to do some thinking about the thinking of computer programs, and to make some evaluative judgments about the evaluative judgments they make.

## REFERENCES

Colby, K. M. Computer simulation of a neurotic process. In S. S. Tomkins & S. Messick (Eds.), *Computer simulation of personality*. New York: Wiley, 1963.

Colby, K. M. Computer simulation of neurotic processes. In R. W. Stacy & B. D. Waxman (Eds.), *Computers in biomedical research.* Vol. 1. New York: Academic Press, 1965.

Colby, K. M., & Gilbert, J. P. Programming a computer model of neurosis. *Journal of Mathematical Psychology,* 1964, vol. 1, 405-417.

deGroot, A. D. *Thought and choice in chess.* New York: Basic Books, 1965.

Feigenbaum, E. A. The simulation of verbal learning behavior. *Proceedings of the Western Joint Computer Conference,* 1961, vol. 19, 121-132.

Green, M., Wahlstrom, S., & Forsen, G. Application of intelligent automata to reconnaissance. Stanford Research Institute Technical Report No. RADC-TR-66-822, May 1967.

Greenblatt, R. D., Eastlake, D. E., & Crocker, S. D. The Greenblatt chess program. *Proceedings of the Fall Joint Computer Conference,* 1967, vol. 31, 801-810.

Loehlin, J. C. The personality of Aldous. *Discovery,* 1962, vol. 23(7), 23-26.

Loehlin, J. C. A computer program that simulates personality. In S. S. Tomkins & S. Messick (Eds.), *Computer simulation of personality.* New York: Wiley, 1963.

Loehlin, J. C. "Interpersonal" experiments with a computer model of personality. *Journal of Personality and Social Psychology,* 1965, vol. 2, 580-584.

Loehlin, J. C. *Computer models of personality.* New York: Random House, 1968. (a)

Loehlin, J. C. Machines with personality. *Science Journal,* 1968, vol. 4(10), 97-101. (b)

Miller, G. A., Galanter, E., & Pribram, K. H. *Plans and the structure of behavior.* New York: Holt, 1960.

Minsky, M. Machines are more than they seem. *Science Journal,* 1968, vol. 4 (10), 3.

Newell, A., Shaw, J. C., & Simon, H. A. Chess-playing programs and the problem of complexity. *IBM Journal of Research and Development,* 1958, vol. 2, 320-335.

Nilsson, N. J., & Raphael, B. Preliminary design of an intelligent robot. In J. T. Tou (Ed.), *Computer and information sciences-II.* New York: Academic Press, 1967.

Rosen, C. A. Machines that act intelligently. *Science Journal,* 1968, vol. 4(10), 108-114.

Samuel, A. L. Some studies in machine learning using the game of checkers. *IBM Journal of Research and Development,* 1959, vol. 3, 211-229.

Samuel, A. L. Some studies in machine learning using the game of checkers. II-Recent progress. *IBM Journal of Research and Development,* 1967, vol. 11, 601-617.

Shannon, C. E. Automatic chess player. *Scientific American,* 1950, vol. 182(2), 48-51.

Simon, H. A. Motivational and emotional controls of cognition. *Psychological Review,* 1967, vol. 74, 29-39.

## DISCUSSION

When the computer program is in its interaction mode, a participant asked Dr. Loehlin, is it technically possible for an optimal path to be charted for this interaction if one starts with any given set of values of the attributes of the two interacting models? Is it possible to discover whether they could interact without engaging in the final hostility output? Is there any path by which they can avoid breaking up the interaction?

Yes, Dr. Loehlin replied, this could be done by determining what kinds of developments preceded the breakdown of the interaction and rerunning the program with intervention introduced at the appropriate point. This could be done by programming in a little loop which would, in effect, instruct one model to change his behavior if the other model does certain things in sequence.

Why is it that computers can perform complex cognitive activities like playing chess and can simulate some fairly complicated personality processes and yet have difficulty walking around the room without bumping into things, another participant asked. Is the inadequacy in the peripheral receptors of the robot or in the central processes?

This is an interesting paradox, Dr. Loehlin said. It seems that artificial intelligence involves working from the hard things to the easy things; or, to put it another way, it is like starting out with the intelligence of adults and working backwards toward the intelligence of children. One possible explanation is that, as adults, we find it easier to understand the adult world and so that is where we start. A more interesting possibility is that perceptual-motor operations that we think are simple because we have been able to do them for a long time are really much harder and more complicated than the kind of planning that goes on in something like playing chess. What the child does during his first year or two in organizing simple muscular movement may actually be more difficult a feat than what he does later when he is building upon what he has done earlier.

How useful is computer programming in predicting physiological adjustment to certain stimuli, a participant inquired. Suppose that we know certain sequences of events take place in the biological organism as a result of the secretion of neuro-hormones and so on. We know the stimulus, and we know the effects. What can we find out about how these physiological adjustments or effects come about?

This could be fairly straightforward, Dr. Loehlin said. If you have several ideas about what the missing links might be, you can program in representations of these ideas. If several of them give you unpromising results and one seems to work, this might be accepted as an hypothesis worth investigating further.

It will be a long time before we can dispense with human and animal study subjects in physiological research, another participant commented. It is possible to run simulations and get results that look promising; yet when these are checked against the living system, the results are not always what the simulation has prepared the investigator to expect. The results you can get with a computer depend on the completeness of the information you are able to feed into it, and there is a great deal we do not know about the physiological constants of animal systems.

It appears that computer devices have certain general properties in common with each other and probably also with brains, a participant said. However, these machines differ from organisms in certain important ways. For one thing, machines can be idle, but brains cannot. A computer stores information and communicates information at the will of the programmer. The reason for its existence is to come up with results. There is an input and an output. When the task is completed, the machine is idle. It stops whatever it was doing. Brains are not really concerned with results, with the completion of specific tasks. They go from stage to stage or state to state, often quite helplessly, with no one to look at results.

True, but the differences may not be quite that extreme, Dr. Loehlin replied. A modern computer does not stand idle, anymore than a brain does. It will stop if you turn off the switch, just as a brain will stop if you shoot it; but it does not stop when it completes a task. It just keeps on computing at something else. It does this because the structure of its operating system makes it do so.

What the motivations of a computer program are depends upon the kind of input it has received and what its structure is. Initially, the programmer puts the motives and purposes into the computer program, but once they are in the program, they are no longer his purposes but the program's purposes, controlled by its structural regulatory devices. The same is true of the brain. Parents may tell a child what to do, but once he incorporates the suggestion into the motive structure of his brain it becomes his purpose, not theirs.

It should be possible to take advantage of the potentially continuous action of these computer programs, a participant suggested. The program could be given residual assignments so that it would go into an exploratory mode when it had nothing else to do. It could be told, when you finish your task, go play checkers, or review the games that are set up over in the corner, or review your interaction with another robot and see what went wrong with it.

Is it not true, a participant asked, that one distinction between a computer and a brain is that both remember the outcome but it is more difficult for the brain to remember the thought processes that led up to the outcome? Are not all of the steps equally accessible in the computer memory? The computer can scan all the data it has stored and find all that is there, but the human brain does not remember everything in a sequence leading to a particular outcome.

It is important to distinguish between the computer, which is a machine, and the computer program as it operates in the machine, Dr. Loehlin said. A computer as a machine may have a perfect memory, but a computer program remembers only what it is designed to remember. Thus it may or may not be able to give a full account of how it arrived at the solution to a problem, depending on what it has been designed to do.

When we start thinking about the computer as something other than a tool, we tend to get threatened by a feeling that human beings are being replaced by machines, a participant said. But the very fact that we find use for it points up the difference. If the experimenter wants to examine a set of processes that occur within the organism, the computer program may be able to do a much better job than the human of looking at variables and arriving at predictions. But it is still the human being who must determine whether the results do in fact represent the kinds of situations that will occur. The computer can only do what the human tells it to do, and it cannot change what it does unless it is told to do so.

This is true up to a point, Dr. Loehlin responded. The computer can be programmed to make certain decisions. The experimenter knows what the computer will do in certain specified contingencies because that is what it is programmed to do, but he does not know all the implications of what he has told it to do. For example, he may tell the computer where to find certain information, and what to do about the information it receives. But he does not know what the information will be; consequently, he does not know what the computer will do next.

The computer program has two separate roles. One is an executive role, in which the computer program is a tool that runs a very complicated experimental show. The other is a model role in which the computer program is a representation of something the experimenter wants to test out. As a model, it does not have to be "smart," any more than the model of a chemical molecule has to be able to enter into a chemical reaction. It just has to represent certain structural or process features that the experimenter builds into it.

It is always fascinating to be able to construct a program that can perform humanlike operations such as playing chess, a participant said, but the important question is, what can these programs tell us about the way humans perform certain operations? What are the similarities between what goes on inside a computer and what goes on inside a person?

Computer programs are theoretical models that do not supply empirical information about how people do things, any more than mathematical models do, Dr. Loehlin replied. But they do indicate what the implications are of the assumptions that have been built into the programs. For example, if you program in a combination of three properties, each of which is a logical one to include, the program will quickly show whether these properties are compatible or whether they are hopelessly contradictory. A mathematical model is very useful for dealing with a small number of variables in a highly controlled experimental situation. However, some problems—particularly in psychology—are much too complex to be solved with mathematical models; and verbal models become far too complicated to be useful. With a computer program, the experimenter can take a complicated configuration of ideas and follow out their implications much more easily.

In fields like weather forecasting and economics, computer modeling has enormous practical value. For example, it is possible to simulate the economy and make detailed predictions about the impact of changing tax rates or interest rates or doing something that will affect employment. Because a lot is known about economic mechanisms, the results can be checked by the computer program at a large number of internal points.

This is much more difficult in an area like psychology, and the practical applications of computer technology in relation to human behavior should not be oversold. It will be very difficult to make reliable predictive simulations, because so much goes on inside the human organism to which we have limited access. But the use of computers is probably preferable to the use of incredibly complicated and confusing verbal models. Using the machine, the psychologist will at least know what it is he thinks instead of only what he thinks he thinks.

This presentation has some very provocative implications for studies of the development of self-regulatory mechanisms in children, a participant said. Our knowledge of self-regulation processes in children is very limited, although some interesting and important work is being done on the ways children give themselves instructions for a subroutine. The process of self-evaluation, in which the child checks what he did against some standard of adequacy that he has established, is under study. As yet, we do not know what intelligence a child needs or what other things he needs to know to be able to make use of his self-regulatory capabilities.

It has always been an interesting unanswered question in child development why a child can discriminate a triangle or a diamond from other shapes at the age of a year but cannot copy it until he is around six. It seemed possible that a younger child draws poorly because he simply operates and does not check back to the model. But a recent series of eye photographs done at Stanford shows that the number of check-backs has no relation either to age or to the quality of the drawing. This makes it appear that the trial, operate, and check processes are internal against a representation and not against an outside model. In this case, research becomes very complicated. What is the self-regulatory process that is involved in this internal check against the internal model?

Impulse control is another area in which much more study is needed. There comes a time when a child reaches a decision point, a branching point where he has to operate a subprogram to get the information he needs to make a choice. This is difficult for children to handle, because they cannot tolerate the tension involved in "sitting it out" while the subprogram is executed before the decision can be made. We are in an abysmal state of ignorance about what is involved in getting that pause to occur. All that has been done so far is to describe the fact that it must occur and to trace age changes in the ease or difficulty with which it occurs. This problem seems to carry some interesting parallels to Dr. Loehlin's description of the "stop and think" instruction the robot must receive so that he will not destroy himself against the walls of the room before he can change his path of movement.

Perhaps the greatest value of Dr. Loehlin's presentation is the new perspective it gives on the state of research on the self-regulatory mechanisms in child development. His listing of some of the feedback mechanisms and other devices that any self-regulatory mechanism has to have makes one ask questions about what mechanisms are used and how they develop in the young human child.

# THE DEVELOPMENT OF SELF-REGULATORY MECHANISMS: EPILOG

*Donald L. Peters*

The Pennsylvania State University

The time has come to survey the major contributions made during this conference, to place them in the larger perspective of early childhood development, and to sort out, at least from a single perspective, those avenues for future research that look particularly promising. Such a task is not easy and, by necessity, it must be biased by one's own knowledge base and interest areas. The attempt, therefore, will be selective rather than inclusive.

First, although it is doubtful that unanimous agreement among the participants could be arrived at, an effort will be made to provide some definitional precision to the notions of self-regulation that have been the central concern of this conference. This will be followed by a look at two major topical themes that pervaded the several presentations and discussions; namely, the dynamic-interactional view of human development and the processes of individualization. Lastly, an effort will be made to summarize those areas for future investigation that have been suggested, with particular emphasis on the possible contributions of interdisciplinary efforts.

## SELF-REGULATORY MECHANISMS

As would be expected, the organizers of this conference made several assumptions as to the nature of the self-regulatory processes of the human organism, about their definition, about the areas of research relevant for their specification, about their development, and about their interrelation. Two major areas, biological regulatory mechanisms and thoughtful (or intellectually mediated) regulatory mechanisms were presupposed to be of particular importance. Since it also was assumed that these areas were more likely to be overlapping than independent in their determination of behavioral outcomes, focus was directed toward the early self-regulatory behaviors of infants, particularly those concerned with responsiveness to the environment, as a point where the interrelation of the two divisions would be most readily apparent. Lastly, since the major emphasis of these meetings was to explicate what is meant by self-regulation and the implications of self-regulation for behavior, the field of computer science was called upon for its clarification of the construct.

What emerged was not an explicit definition of "self-regulation" nor a taxonomic listing of behavioral, physiological, or psychological occurrences which delimit the realm of the phenomena in any finite way. Rather, there evolved a sense that self-regulation in the organism is best described as a dynamic process of adaptive functioning operating through a large number of individually distinct mechanisms which may be hierarchically arranged or classified. That is, whether one looks at self-regulation at the cellular level as mentioned by Dr. Rosenzweig or at the individual adaptation of communications to the needs of the listener as discussed by Dr. Krauss, one witnesses a dynamic process with adaptive change as its outcome. To reiterate Dr. Inhelder's own terms:

> . . . the regulatory systems are found on all levels of the organism's functioning, right from the genome up to psychological behaviors; thus they appear to be among the most general characteristics of the organism. Self-regulations seem to constitute at the same time one of the most universal characteristics of life and the most general mechanisms to be found in both organic and cognitive behaviors.

The self part of self-regulation derives from the notion that the mechanisms of change exist within the organism and that the resulting change is observable in the organism itself or in its behavior. The stimulus for the activation of a particular mechanism may be internal or external, but it generally involves a discrepancy, imbalance, or disequilibrium. The complexity of the mechanism involved may range from the most simple to the most complex with complexity, borrowing Dr. Loehlin's terminology, referring to the number of conditional branches, subroutines, and exits that are possible. The resulting change may be homeostatic—a return to an optimal level of functioning, or it may be progressive—a movement to a new, presumably more adaptive level. In either case a feedback loop is implied.

Biological regulatory mechanisms refer primarily to biochemical or biophysical regulations which govern the organization, structure, or growth of the organism. Thoughtful self-regulation refers primarily to those regulations which reflect the mediation of sensory-motor or symbolic operations. The latter may or may not reflect conscious intentionality. Both types are likely to have behavioral manifestations.

The distinction between biological and thoughtful regulatory mechanisms may be made another way. In a sense, the biological regulatory mechanisms explored by Rosenzweig and Denenberg are in the realm of intervening variables between the stimulus experience and the behavioral outcomes. Each of these investigators has pursued, through a systematic series of controlled experiments, the neuro-physiological or neuroendocrinological mediators of behavior changes resulting from the manipulation of early experience. That is, they have sought mediators which have concrete existence.

Dr. Steinschneider was also inquiring into the possibility of intervening variables when he suggested that cardiac change may itself be the endpoint of a series of biochemical steps between the stimulus and behavior. However, with his paper and Dr. McCall's there is a shift to the realm of hypothetical constructs. From this point on, with the exception of Lenneberg's discussion of the brain correlates of language, concern is with the psychological area of hypothesized cognitive structures without direct physiological or biochemical referents.

From this distinction, and from the evidence presented by the conference speakers, it appears that a temporal ordering of the development of self-regulation is possible. Both Rosenzweig and Denenberg cite examples of the ontogeny of "brain organization" occurring early in the lives of their animals. Dr. Denenberg's examples of the key role of the hormones testosterone and corticosterone in the organization of the rat brain are most convincing. It seems likely that these predate those regulatory mechanisms involving alterations of cognitive structures. Indeed, they appear to predate the changes in the cerebral cortex reported by Rosenzweig. The cognitive adaptations of the child in the sensorimotor period, and the habituation to visual and auditory stimuli discussed by Dr. McCall appear next in the chronology followed by the "conscious" social adaptations of Dr. Krauss' older subjects.

Although the origins of particular self-regulatory mechanisms may be traced chronologically, it is likely that in the functioning organism many different levels of self-regulation operate concurrently. As analysis of the mediators of behavior proceeds from the micro level to the macro level, a pyramid of self-regulatory mechanisms, all functioning simultaneously, is encountered with each flowing toward the final behavioral product. Analysis at all levels, and across levels is needed to provide better understanding of the child and adult.

## THE DYNAMIC INTERACTIONAL VIEW OF HUMAN DEVELOPMENT

Much of the discussion of this conference focused upon the development of the organism as a function of its dynamic interactions with its environment (physical or social) after birth. This, no doubt, reflects the general acceptance by the participants of the importance of experiential factors in development, whether mediated through learning or through physiological processes.

The focus on the postbirth period should be viewed as a choice which does not imply a lack of concern for prenatal factors that may also affect the development of self-regulatory mechanisms. Indeed, the participants on several occasions referred to influences during the period between conception and birth as being significant to the understanding of later development. The discussions suggest the importance of tracing the origins of early biological self-regulatory mechanism to the prenatal period. Particularly when seeking the antecedents of inadequate or inappropriate development of self-regulation the prenatal period is a fruitful one to investigate. The extensive researches of Benjamin Pasamanick and Hilda Knobloch and their associates e.g., (see Pasamanick & Knobloch, 1966) point to a wide range of potentially important environmental factors and their behavioral manifestations. Nor should prenatal nutritional influences be neglected (Hurley, 1968; Tompkins, 1948).

It should also be noted that conception itself is an arbitrary point of departure for understanding the individual. It would have been equally appropriate to have focused the discussions on some time prior to conception. There is evidence to support the powerful, if somewhat indirect, influence on the subsequent development of the child of such factors as individual and social constraints on mate selection, the preconceptual emotional and physiological readiness of the female, and the family's and society's acceptance of the pregnancy (Berrill, 1968). Each of these influence the mother's acceptance of the role of motherhood, her capacity for bearing healthy children, her subsequent relations with the child, and, hence, the very nature of the new person she will bear and rear. Denenberg has provided some empirical evidence of these long range, indirect environmental influences (Denenberg & Rosenberg, 1967).

To cite these preconceptual, prenatal, and postnatal influences on the behavior and development of the child does not provide the total perspective of the development of self-regulation. As evidenced in the papers of Drs. Steinschneider, McCall, and Inhelder, it is equally important to consider the activity of the child himself. The active, exploring nature of the child, his persistent, but selective, seeking of stimulation yields important fruits for his own development. This activity, gaged to the organism's present level, provides the inputs needed for the next step of development. Infants seek out stimulation and expend effort to perform behaviors that are their own reward. Both Drs. Rosenzweig and Inhelder have pointed to play as a manifestation of this process.

It would seem that the activity of the child is important to his own development for at least two related reasons. As suggested by Inhelder it is through such activity that the child develops his concepts of space, time, and causality. Through the gradual processes of assimilation and accommodations his cognitive structures are developed to approximations of those of the adult. The second reason, pointed to by both Dr. Lipsitt and Dr. Inhelder, is that through these early learning activities the child has the opportunity to see that "what he does makes a difference." Through the sensory feedback he learns to adjust to his world, and adjust his world to him.

## THE PROCESSES OF INDIVIDUALIZATION

Concern for the importance of the activity of the young child leads directly to the next point that was evident throughout the conference proceedings. When considering the development of self-regulatory mechanisms it is necessary to consider the origins and importance of individual differences. Some of the commonality of this concern can be demonstrated by recalling Rosenzweig's discussion of the "Nonspecific Excitatory Level" (NEL) in the rat and its relation to dietary preferences and Denenberg's inclusion of emotionality and exploration as two factors affected by infant handling. The research of both men can be construed as being concerned with the origins of individual differences. One can also recall Steinschneider's statement that,

> The neonate is not a passive recipient of environmental inputs. Rather, he responds to environmental demands in an individually characteristic manner which has the potential for modifying the interactional process.

One should also note Steinschneider's explication of the importance of the individual's cross-over point (the initial heart rate value at which no heart rate change occurs with stimulation) as a potential index of individual differences. McCall's emphasis on "short lookers," "rapid habituators," and "slow habituators" provides another view of individualization. Many other instances could be cited.

The above references echo a recurring theme in the study of young children (Birns, 1965; Chess, Thomas, & Birch, 1967; Escalona, 1968; Escalona & Heider, 1959; Schaefer & Bayley, 1963). Normal infants differ from one another in excitability, activity level, and sensitivity to stimulation and these differences are relatively stable over time, at least during the early years.

The importance of these individual differences lies not in their stability, but in their effect at the time. Although there is some evidence that with increasing age there is a movement toward homogeneity in activity level (Escalona, 1968,

p. 23) the value of measuring these early differences derives from the contention that the child's activity is a codeterminer of subsequent development. If the child has a low level of interaction with his environment during the first year of his life, the developmental patterns formed during this year would be partially determined by this level of activity. As such they would be quite different than those of a highly active child in the same environment. If the low activity infant later shifted to a higher level of activity his development would reflect his new activity level, but, since later development builds upon what came before, such subsequent development would also bear the stamp of his previous low level of activity. The importance of this to the area of cognitive development becomes clear in Dr. Inhelder's paper.

In addition to varying in activity level, the majority of infants react to some varieties of stimulation more than to others. The research reported by McCall points this out most explicitly. His explorations of the effects of auditory, visual, and manipulative interaction with stimuli suggest preferences for modalities which are sex-linked, but may also reflect individual differences of considerable importance. Escalona (1968) has carefully researched the value of such preferences in the relation of the child with the mother. She focuses upon the possible convergence or divergence between the child's preferred modality and the modality emphasized by the mother in her contact with the child. An infant highly sensitive to touch whose mother seldom handles him is an example of divergence. The modality to which the infant is highly reactive is evoked minimally and hence this preferred pathway for learning is little used. The infant who is far more reactive to auditory stimuli than to vision, and whose mother emphasizes speech and sounds provides an example of convergence which presumably fosters the development of the child.

Within the field of education the importance of individual differences has been stressed (Cronbach, 1957, 1967; Gagne, 1967; DiVesta, Peters, Sanders, Shultz, & Weener, 1970) and the implications for early childhood education have begun to be explored (Beller, 1967; Peters, 1967, 1970). However, far too little is yet known about the origins of individual differences in activity level and modality performances. Lacking also is knowledge of the full range of ramifications of such differences for learning and interaction with the environment during the very early years of life. Research such as reported by the participants of this conference, with its stress on the importance of early experience, should provide a good foundation for efforts which seek to fill the gaps in our knowledge.

## RESEARCH IMPLICATIONS

At the end of a conference such as this, the next question is, "What are the implications for future research?" Each of the formal presentations and many of the comments during the discussions have suggested routes for profitable inquiry, specific problems needing analysis, and necessary methodological refinements. The details will not be repeated here. Rather, what follows is an attempt to conceptualize three approaches to the research questions that have been raised. All three are based upon a faith in the gradual accretion of knowledge through systematic research and a concern for the social relevance of knowledge. No one is construed as having priority over the others and, in fact, it would seem that all three should be pursued with vigor.

The first could be called the Unitary Problem Approach. This is not the traditional single discipline approach but one that involves an interdisciplinary team of researchers focusing their talents on the task of providing an ever finer understanding of the variables of a particular problem area. Here the concern is with the depth rather than breadth of understanding, but always with a cognizance of and commitment to the value of unifying the focus of several disciplines.

In the research reported at this conference there have been several outstanding examples of the unitary approach. They may well serve as models for others. Refinement of the construct of attention, the relationship of the orienting response to cardiac and biochemical indices, computer simulation of biological and psychological models, and the relationships between brain organization, brain function, and language knowledge are all likely to yield new insights under systematic applications of this approach.

The second approach could be considered an Integrative Approach. This requires seeking and understanding in a wider context, not only across disciplines but across problems. In terms of the conference format it would require building unique bridges across sessions. For example, the concern would not only be to investigate the biochemical mediators of emotionality, but to extend beyond to both other physiological variables and to other behavioral indices, possibly in the realm of social or maternal behaviors. The Integrative Approach would also concern itself with finding continuities and discontinuities across the life span and across species.

The Integrative Approach would seem to be required to investigate the kind of multivariable, multilevel situation found when one attempts to investigate the self-regulatory mechanisms of human beings in social communication situations. For example, analysis of the sort described by Dr. Krauss probes only one of the important dimensions of communications. Nonverbal communications at the macro and micro levels would also be important. Kinesics, personal space relationships, person perception, auditory sensitivity, olfactory sensitivity, environmental constraints, and many other variables could be considered. At another level there is concern for the nature of language itself as expressed by Dr. Lenneberg. Each of these areas indicates a set of interrelated problems generally to do with communications.

The third research approach is the Implementation Approach. This approach implies bringing knowledge to bear in the direct solution of social problems. In methodology it can overlap with either of the other approaches, but it differs from them in its goal. The goal in this case is social change, or intervention. It frequently assumes, or accepts, the results of prior empirical research and seeks to apply this knowledge in field settings, in communities, with individuals or with groups. For example, research on early experience has been applied in such intervention programs as Head Start, and research on learning has been applied in public education. When this approach is undertaken with appropriate rigor it can be an effective source of knowledge as well as an initiator of social change. The danger lies in the possible loss of the word "research" from the phrase "Implementation Approach for Research."

Each of the three approaches discussed is more difficult than the traditional single discipline approach. Interdisciplinary research is fraught with communication difficulties and is frequently considered outside the normal reward systems of a profession. The problems become increasingly complex as one approaches the Integrative and Implementation approaches. However, the potential payoffs for such efforts in terms of new knowledge and in the building of broad theories, appear to be of such magnitude as to warrant the extra effort and risk.

## REFERENCES

Beller, E. Methods of language training and cognitive styles in lower class children. Paper presented at the annual meeting of the American Education and Research Association, New York, February 1967.

Berrill, N. *The person in the womb,* New York: Dodd, Mead, 1968.

Birns, B. Individual differences in human neonates; responses to stimulation. *Child Development,* 1965, vol. 36, 249-256.

Chess, S., Thomas, A., & Birch, H. Behavior problems revisited. *American Academy of Child Psychiatry Journal,* 1967, vol. 6, 321-331.

Cronbach, L. The two disciplines of scientific psychology. *American Psychologist,* 1957, vol. 12, 671-684.

Cronbach, L. How instruction can be adapted to individual differences. R. Gagné (Ed.), *Learning and individual differences,* Columbus, Ohio: Merrill, 1967.

Denenberg, V., & Rosenberg, K. Nongenetic transmission of information. *Nature (London),* 1967, vol. 216, 549-550.

DiVesta, F., Peters, D., Sanders, N., Shultz, C., & Weener, P. Instructional strategies: Multivariable studies of psychological processes related to instruction. Semi-Annual Report, Advanced Research Projects Agency (ARPA #1269), January 1970.

Escalona, S. *The roots of individuality,* Chicago: Aldine, 1968.

Escalona, S., & Heider, G. *Prediction and outcome: A study of child development.* New York: Basic Books, 1959.

Gagné, R. *Learning and individual differences,* Columbus, Ohio: Merrill, 1967.

Hurley, L. The consequences of fetal impoverishment. *Nutrition Today,* 1968, vol. 3, 3-9.

Pasamanick, B., & Knobloch, H. Retrospective studies on the epidemiology of reproductive casualty: Old and new. *Merrill-Palmer Quarterly,* 1966, vol. 12, 7-26.

Peters, D. L. Task variations and individual differences in Piaget's conservation of number. *Merrill-Palmer Quarterly,* 1967, vol. 13, 243-258.

Peters, D. L. Verbal mediators and cue discrimination in the transition from nonconservation to conservation of number. *Child Development,* 1970, vol. 41, 707-721.

Schaefer, E., & Bayley, N. Maternal behavior, child behavior, and their intercorrelations from infancy through adolescence. *Monographs of the Society for Research in Child Development,* 1963, vol. 28, 3.

Tompkins, W. The clinical significance of nutritional deficiencies in pregnancy. *Bulletin of New York Academy of Medicine,* 1948, vol. 24, 376-388.

# AUTHOR INDEX


A

Ader, R., 56, *62*
Alexander, A. A., 81, *102*
Ames, E. W., 112, 113, 114, 117, *133, 136*
Anokhin, P. K., 17, *31*
Arnold, W. J., 21, *33*
Atkinson, J. W., 117, *135*

B

Bartels, B., 93, 95, 97, *101*, 117, 128, 129, *135*
Barten, S., 93, *100*
Bayley, N., 131, 132, *133*, 235, *239*
Beintema, D., 94, *101*
Belfer, M. S., 56, *62*
Beller, E., 236, *238*
Bellugi, U., 174, *183*
Benacerraf, P., 160, *183*
Bennett, E. L., 24, 25, 26, 27, *31, 32, 33*
Bennett, T. L., 26, *31*
Berlyne, D. E., 111, *133*
Berrill, N., 234, *238*
Bever, J., 99, 100, *101, 102*
Birch, H., 235, *238*
Birns, B., 93, *100,* 235, *238*
Blatchley, M. E., 132, *136*
Block, J. D., 87, *100*
Bolles, R. C., 18, 19, *31*
Bremermann, H., 15, 16, 17, *31*
Brennan, W. L., 112, 113, 114, *133*
Bridger, W. H., 76, 87, 89, 93, 98, *100, 101*
Brown, R., 188, 189, 200, *202*
Brumaghim, J. T., 53, 54, *63, 64*
Buneman, O. P., 169, *184*

C

Cameron. J.. 131, 132, *133*
Campbell, H. J., 61, *62*,117, 128, 129, *135*
Cantor, G. N., 117, 129, 131, *136*
Caron, A. J., 128, *133*
Caron, R. F., 128, *133*
Charlesworth, W. R., 115, *133*
Cheng, M.-F., 23, 24, *33*
Chess, S., 235, *238*
Chevalier, J. A., 40, *63*
Chun, B. J., 74, *101*

Clark, R. A., 117, *135*
Clay, J., 21, *32*
Clifton, R. K., 74, 98, 100, *101,* 109, 134
Cohen, B. D., 195, 198, *202*
Cohen, L. B., 109, 111, 114, 116, 119, 128, 131, *133, 136*
Colby, K. M., 215, *225, 226*
Collins, D., 110, 111, *134*
Conklin, P. M., 56, *62*
Cornwell, A. C., 20, *33*
Coursin, D. B., 111, *134*
Cragg, B. G., 26, *31*
Crocker, S. D., 218, *226*
Cronbach, L., 236, *238*
Cronin, V., 93, *100*
Crowell, D. H., 74, *101*
Curcio, F., 116, 128, *133*

D

Davis, C. M., 74, *101*
DeBoer, K., 99, *102*
deGroot, A. D., 218, *226*
Dember, W. N., 117, 128, *133*
Denenberg, V. H., 27, *31*, 40, 42, 43, 44, 45, 46, 47, 48, 50, 51, 52, 53, 54, 56, 57, 58, 59, 60, 61, *62, 63, 64,* 132, *136*, 234, *238*
Diamond, M. C., 24, 25, *31, 33*
DiVesta, F., 236, *238*
Dodd, C., 93, 95, 97, *101*

E

Earl, R. W., 117, 128, *133*
Eastlake, D. E., 218, *226*
Eayrs, J. T., 27, *31*
Eiduson, S., 52, *64*
Eisenberg, R. B., 111, *134*
Ellis, H. C., 26, *31*
Escalona, S., 235, 236, *238*

F

Fadel, D., 117, 128, 129, *135*
Fantz, R. L., 111, 114, 115, 116, 128, 132, *134*
Feigenbaum, E. A., 224, *226*
Fitzgerald, H. E., 110, *134*
Flavell, J. H., 192, *202*
Forsen, G., 221, *226*
Fortier, C., 54, *63*

G

Gagné, R., 236, *238*
Galanter, E., 211, 212, *226*
Garcia, J., 22, *31*
Gardner, B. T., 173, *183*
Gardner, R. A., 173, *183*
Garratt, C. R., 110. 130, *135*
Geller, E., 52, *64*
Germana, J., 100, *101*
Gibson, E. J., 26, *31*
Gilbert, J. P., 215, *226*
Glucksberg, S., 191, 194, 195, 196, 198, *202*
Glydon, R., 61, *63*
Goetzinger, C. P., 99, *101*
Goldberg, S., 93, 95, 97, *101,* 128, 130, *134, 135*
Gollová-Hémon, E., 22, *32*
Gottlieb, G., 18, *32*
Goy, R. W., 48, *64*
Graham, F. K., 74, 80, 97, 98, 100, *101,* 109, *134*
Green, M., 221, *226*
Greenblatt, R. D., 218, *226*
Greenfield, N. S., 81, *102*
Griffin, E. J., 111, *134*
Gullickson, G. R., 100, *101*
Guttman, N., 176, *183*

H

Haith, M. M., 110, 111, *134*
Haltmeyer, G. C., 44, 45, 47, 50, 51, 52, 53, 54, *62, 64*
Hargreaves, F., 21, *32*
Harris, L. J., 21, *32*
Harwitz, M., 93, 95, 97, *101*
Hebb, D. O., 40, *64,* 117, *134*
Heider, G., 235, *238*
Held, R., 130, *134*
Henker, B. A., 109, 110, *134*
Hen-Tov, A., 109, 110, *134*
Hershenson, M., 111, 113, *134*

Hess, J. L., 46, 48, *63*
Hnatiow, M., 100, *101*
Hockett, C. F., 170, *183*
Honzik, M. P., 20, *32*
Hord, D. J., 81, *101*
Horowitz, F. D., 114, 131, *134*
Hubel, D. H., 17, *32*
Hunt, E. L., 22, *31*
Hunt, H. F., 40, *63*
Hunt, J. McV., 107, 108, 116, 117, 128, *134, 136*
Hunter, M. A., 111, *134*
Hurley, L., 234, *238*

I

Inhelder, B., *155*

J

Jackson, J. C., 80, 97, 98, *101*
James, T. N., 99, *101*
Jeffrey, W. E., 108, *134*
Johnson, L. C., 81, *101*

K

Kagan, J., 74, *101,* 109, 110, 112, 118, 120, 121, 123, 124, 125, 129, 130, 131, *134, 135*
Kalafat, J., 109, *135*
Kaplan, B., 192, *203*
Karas, G. G., 44, 45, 47, *63*
Karmel, B. Z., 112, 113, 114, *135*
Keesey, R. E., 23, *32*
Kennedy, G. C., 17, *32*
Kerpelman, L. C., 26, *32*
Kessen, W., 110, 111, 128, *134, 136*
Kimeldorf, D. J., 22, *31*
Klein, J. F., 195, 198, *202*
Klein, S. B., 100, *101*
Knobloch, H., 234, *239*
Korchin, S. J., 40, *63*
Kotses, H., 99, 100, *101, 102*
Krauss, R. M., 188, 191, 192, 194, 195, 196, 198, 200, *202, 203*
Křeček, J., 19, *32*
Krech, D., 25, 26, *32, 33*

L

Lacey, B. C., 74, *101,* 109, *135*
Lacey, J. I., 74, *101,* 109, *135*
Lang, P. J., 100, *101*
Lantz, D. L., 200, *203*
Lát, J., 22, *32*
La Torre, J. C., 27, *32*
Leiman, A. L., 24, *33*
Le Magnen, J., 21, *32*
Lenard, H. G., 92, *101*
Lenneberg, E. H., 159, 170, 175, *183,* 189, 200, *202*
Lessac, M. S., 17, *32*
Levine, J., 109, 110, *134*
Levine, S., 19, 21, *33,* 40, 42, 43, 44, 45, 47, 48, 49, *63*
Lewis, M., 93, 95, 97, *101,* 109, 110, 117, 128, 129, 130, *134, 135*
Lézine, I., 150, *155*
Lipton, E. L., 75, 77, 78, 79, 80, 82, 84, 85, 87, *101, 102*
Livson, N., 131, 132, *133*
Loehlin, J. C., 214, 225, *226*
Lowell, E. L., 117, *135*

M

McCall, R. B., 110, 112, 117, 118, 119, 120, 121, 123, 124, 125, 127, 128, 129, 130, 131, *135*
McClelland, D. C., 117, *135*
McDonald, D. G., 81, *101*
McEwen, B. S., 56, 57, *63*
McKee, J. P., 20, *32*
Malkasian, D. R., 27, 28, 29, *32*
Melson, W. H., 110, 112, 117, 120, 123, 124, 125, 127, 129, *135*
Meyers, W. J., 100, *101,* 117, 129, 131, *136*
Miller, G. A., 211, 212, *226*
Minsky, M., 211, *226*
Moffett, A., 112, 117, *136*
Moore, R. W., 112, 113, 114, *133*
Moore, T., 131, 132, *136*
Moss, H. A., 74, *101,* 109, *135*
Mullins, R. F., Jr., 48, *63*
Munsinger, H., 111, 114, 128, *134, 136*

N

Nevis, S., 128, 132, *134*
Newell, A., 220, *226*
Newton, G., 42, *63*
Newton, L., 93, *100*
Nilsson, N. J., 221, *226*
Noirot, E., 18, *32*
Nováková, V., 19, *32*

O

O'Connor, W. B., 56, 60, *64*
Ordy, J. M., 115, *134*
Otis, L. S., 40, *63*
Ottinger, D. R., 132, *136*

P

Paden, L. Y., 114, 131, *134*
Pancratz, C., 131, *136*
Parry, M. H., 110, *136*
Pasamanick, B., 234, *239*
Peters, D. L., 236, *238, 239*
Pfeifer, W. D., 46, 48, *63*
Pfungst, O., 173, *183*
Philpott, J. E., 56, 58, 59, 60, 61, *64*
Phoenix, C. H., 48, *64*
Piaget, J., 108, 128, 130, *136,* 142, 143, 145, 146, 147, 148, 149, 153, 154, *155,* 189, 195, *203*
Poggiani, 20, *33*
Poole, R., 99, *101*
Powley, T. L., 23, *32*
Pratt, R. J., 116, 119, *136*
Prechtl, H. F. R., 92, 94, *101*
Premack, D., 175, *184*
Pribram, K. H., 211, 212, *226*
Putnam, H., 160, *183*
Pye, D., 18, *32*

R

Raphael, B., 221, *226*
Raskin, D. C., 99, 100, *101, 102*
Rausch, M., 128, *135*
Reeve, R., 99, *102*
Riege, W. H., 26, *33*
Reiser, M. F., 76, 89, 98, *101*
Richmond, J. B., 75, 78, 79, 80, 82, 84, 85, *101, 102*
Roessler, R., 81, *102*
Rosen, C. A., 221, *226*
Rosenberg, K., 234, *238*
Rosenblatt, J. S., 20, *33*
Rosenzweig, M. R., 24, 25, 26, 27, 28, 29, *31, 32, 33*
Rousey, C. L., 99, *101, 102*
Rozin, P., 21, 22, 23, 24, *33*
Rubenstein, J., 132, *136*
Rzóska, J., 22, *33*

S

Saayman, G., 117, *136*
Saffran, E., 25, *33*
Salapatek, P., 111, 112, *136*
Samuel, A. L., 220, *226*
Sanders, N., 236, *238*
Sandritter, W., 19, *32*
Schaefer, E., 235, *239*
Schaffer, H. R., 110, *136*
Schapiro, S., 52, *64*
Schlutter, G., 19, *32*
Schwartz, L. S., 56, 57, *63*
Sheard, N. M., 18, *33*
Shannon, C. E., 218, *226*
Shaw, J. C., 220, *226*
Sheldon, A. B., 119, *136*
Shultz, C., 236, *238*
Simon, H. A., 213, 220, *226*
Sinclair, H., *155*
Siqueland, E. R., 110, 116, *136*
Slagle, R. W., 25, *33*
Smith, D. B. D., 99, 100, *102*
Snyder, C., 99, *102*
Sokolov, Y. N., 100, *102*, 109, *136*
Solomon, R. L., 17, *32*
Spears, W. C., 111, *136*
Spellacy, F. J., 74, *101*
Spencer, W. A., 128, *136*
Stambak, M., 150, *155*
Stanley, W. C., 20, *33*
Stefflre, V., 200, *203*

Strawbridge, P. J., 99, 100, *102*
Steinschneider, A., 75, 77, 78, 79, 80, 82, 84, 85, 87, 99, *101, 102*

T

Teitelbaum, P., 23, 24, *33*
Thatcher, J., 50, 51, 52, 53, *63, 64*
Thoman, E. B., 19, 21, *33*
Thomas, A., 235, *238*
Thompson, R. F., 128, *136*
Thompson, W. R., 56, *64*
Tighe, T. J., 26, *31*
Tompkins, W., 234, *239*
Trattner, A., 20, *33*
Tucker, L. A., 119, *135*

U

Udelf, M. S., 115, *134*
Uzgiris, I. C., 116, *136*

V

Vivekananthan, P. S., 188, 192, 200, *202*
von Bernuth, H., 92, *101*

W

Wade, G. N., 22, *33*
Wahlstrom, S., 221, *226*
Walk, R. D., 26, *31*
Ward, A., 21, *32*
Watson, J. S., 130, *136*
Weener, P., 236, *238*
Weiffenbach, J. M., 119, *135*
Weinheimer, S., 188, 192, 200, *202, 203*
Weir, W., 114, *136*
Weisberg, R., 191, 195, *202*
Weiss, J. M., 56, 57, *63*
Weizmann, F., 116, 119, *136*
Werner, H., 192, *203*
Wetzel, A., 19, 21, *33*
Whimbey, A. E., 42, 43, *64*
Whitaker, H. A., 176, *184*
White, B. L., 132, *136*
Wiesner, B.P., 18, *33*
Wilder, J., 86, *102*
Woods, J. B., 18, 19, *31*
Wu, S-Y., 25, *33*

Y

Young, W. C., 48, *64*

Z

Zarrow, M. X., 27, *31,* 44, 46, 48, 50, 51, 52, 53, 54, 56, 58, 59, 60, 61, *63, 64*
Zeeman, E. C., 169, *184*
Zucker, I., 22, *33*

# SUBJECT INDEX


A

Accommodation, 108, 132, 150, 235
Adaptability, 101
Adaptation, 34, 232, 233
biological, 142
cognitive, 153
to stress, 47
to testing, 34
Adrenal cortex, 43
Adreno-cortical activity 3, 31, 43, 45–47, 52, 54, 64, 65
Adreno-cortical response, 4, 46
Aggression, 42
Aldous, 213, 214, 226
Anger, 1, 11
Anxiety, 101
Assimilation, 3, 8, 9, 108, 142, 147, 150, 235
reciprocal, 147
Attachment, mother/child, 71
Attention, 107
auditory, 101, 124, 131, 137
in computers, 224
definition of, 107, 109, 139
differential, 108, 132
fixation
first, 109, 110, 122–124, 126, 131, 139
total, 109
habituation of, 8, 121, 128, 129, 131–133, 137
latency, 109
measurement of, 109–111
pupil dilation, 110
visual, 107, 109–112, 131
Autonomic nervous system, 73
Autonomic response patterns, 135
system, 74, 101
Avoidance
conditioning of, 4, 22
learning, 22, 40, 42, 63, 66

B

Behavioral state, effect on CRR, 74, 75, 92–98, 100–101
Bradycardia, 100
Brain
chemistry, 3, 24
correlates of language, 176, 183, 184
functions, 158, 177, 178

organization of, 16, 28, 34, 36, 62, 63, 65, 66, 233
plasticity of, 17, 178
structure, 16, 21, 34, 36, 37, 158, 228
weight, 3, 24, 25

C

Caloric regulation, 21
Cardiac response
acceleration of, 6, 7, 74, 83–85, 92, 93, 96, 98, 99, 101, 103, 104
deceleration of, 6, 7, 74, 83–85, 88–93, 96–98, 101, 103, 104, 109, 110, 118, 124–127, 129
latency, 82
peak magnitude, 6, 79–81, 84–86, 97, 110
rate of, 6, 7, 74–76, 81, 82, 85, 91, 92, 98, 129, 136, 138, 233
return magnitude, 79–82, 88–90
Central nervous system, 4, 15, 143, 154
activation of, 48, 62
organization of, 48, 49, 55, 56
Cerebral cortex, 24, 25
Chomsky, N., 141
Cold stress, 53
Communications, 9, 10, 11, 167, 169, 171, 178, 180, 185, 188, 189, 191–200, 202, 204–208, 232, 237
communality of, 10, 207
communality index, 200
with computers, 227
effectiveness of, 10
regulation of, 11
Computer models, 3, 209–226
programs, 11, 12
Computer simulation of decision making, 11, 12
Conditional branch, 210, 221, 232
Conflict, 216
Corticosterone, 4, 5, 43–47, 49, 50, 52–58, 60–65, 233
Corticotropic releasing factor (CRF), 61, 65
Critical periods, 48, 62, 68, 70
Crossover point, 89–92, 235
Curiosity, 12, 233
curiosity behavior, 133
Cybernetics, 142, 154

D

Decision instruction, 210
Defecation, 42, 45
Defense reflex, 7, 74, 100, 101, 104
Dendritic spines, 25
Deprivation, social, 33, 37
Discrepancy, 115–118, 121, 123–125, 129, 232
information in computers, 211
magnitude of, 117, 118, 123, 125, 127, 135, 137
parameters of, 116
Discrimination learning, 42

E

Electrocardiogram, 76
Emotional reactivity
definition of, 42
in rats, 1, 42, 43, 45, 47, 62, 63
Emotionality, 4, 43–45, 56, 64, 66, 214–216, 218, 235
Endocrine function, 32
Environment
cardiac response to, 5
complexity of, 33
enriched, effect of, 17, 24–26, 28, 29, 31, 32, 34, 35, 40, 66, 215, 234
impoverished, effect of, 24, 25, 26, 29, 69, 215, 234
novelty of, 44, 46
Estrogen, 4, 48, 50
effects on body weight, 22, 23
on sex behavior, 22, 23
in food intake, 22, 23
Ethology, 16
Exploratory behavior
in infants, 133, 135
in rats, 34, 35, 42, 43, 62, 66, 234, 235

F

Familiar stimuli, 108, 115–117, 138, 139
Fear, 1, 11, 24
Feedback, 215, 216, 232, 235
mechanisms of, 154

sensory, perceptual, 70
Field exploration in rats, 43
Food intake
control of, 3, 15, 17–21
effects of estrogen on, 22
of thyroidectomyon, 23, 24

G

Galvanic skin response (GSR), 74, 100
Genetics, 15, 16, 30, 99, 142, 154
mathematical aspects of, 16
Glia, 25, 36
Grammar, 185, 186, 202, 204
Growth, 67
rate in rats, 18

H

Habituation, 8, 45
pattern of, 123, 125, 127
rapid, 8, 116, 122, 124, 125, 132, 235
to stimuli, 100, 103
to visual stimuli, 108, 117, 118
Habituators, slow, 8, 122, 124, 125, 235
Handling, 4, 28, 31, 39–41, 235
definition of, 41
effects of, 42–47
Harvey, William, 20
Heat stress, 50, 52–54, 57
definition, 50
effects of, 50
Hebb–Williams maze, 3, 4, 21, 66
Hippocampus, 57
Holophrases, 148
Hypothalamus, 3, 5, 23, 24, 33, 43, 50, 55–58, 61, 62, 65
lesions of, 18

I

Imagery, 192
Imitation, 148–150
Impulse control
in children, 1, 2
simulation, 12
Infant vocalization, 132, 133
Intrauterine experience, 99
Isolation, effects of
in puppies, 17
in rats, 3, 29, 32
Iterative loop, 211–213

L

Language, 148–150
acquisition of, 142
characteristics of, 153
components of, 165
natural, 159, 161, 164–168, 171, 173, 185
structure of, 10
synthetic, 166
teaching of, 10
theories of, 141
use of, 148
Language knowledge, 2, 9–11, 157, 167, 170, 172, 184, 187
Law of initial value, 86, 87, 89
Linguistics, 187
competence in, 141, 142

M

Memory, 16, 26, 31, 107, 108, 115, 116, 128, 132, 170, 184
in computers, 210, 214, 222, 228
retrieval of, 108

N

Neural stressors, 54
Neuroendocrinology, mechanisms of, 2, 3, 31
Neurophysiology, 2, 3
Nonspecific excitatory level (NEL), 22, 34, 235
Novel stimuli, 108, 116, 135, 139, 191
effects of novelty, 4, 44, 46
in environment, 46
Nutrition, 3, 5, 32, 34, 35, 37, 67, 235

O

Object permanency, 115, 143, 145
Olfactory discrimination, 20, 68
Openfield activity, 42, 57, 63, 64, 66
Orienting reflex, 7, 74, 100, 101, 103, 104, 109, 134

P

Peak magnitude (in cardiac response), 6, 79–81, 84–86, 97, 110
Perceptual learning, 108
Piaget J, 143, 145–148, 153–155, 181, 208, 213, 223
Pituitary, 43, 58, 59, 61, 63, 64
  growth hormone, 23
Play
  in infants, 134
  integrated, 110
  instinct, 2, 16
  parallel, 110
  symbolic, 142, 149
Postweaning period, 24
Problem solving, 4, 34, 220
  in rats, 34
Progesterone, 48, 50

R

Recursion, 211–213
Reflex sucking, 20, 21
Relation-space in language, 169, 184
Relations in language, 158, 164, 165, 168, 170, 173, 175, 176, 179–181, 185
Releasers
  of behavior, 16
  releasing factor, 65
Respiratory response, 99
RNA, 19

S

Secondary circular responses, 110
Self organization, 2, 3
Semantics, 165, 185
Sensorimotor behavior, 142–153, 223
Sensorimotor intelligence, 142
Sensorimotor period, 8, 139, 143, 145, 147, 233
Sex differences, 101, 236
  in attention, 119, 130, 131, 133, 134, 139
  in cardiac response, 100
Sokolo V, 100, 109
Speech, 10
Stimulation
  in infants, 40, 49, 62
  novel, 108, 116, 135, 139
  social, in rats, 26, 28, 37
Stimulus
  contour, 111–115, 135
  discrepancy, 115–118, 121, 123–125, 129
  familiar, 108, 115–117, 138, 139
Stimulus complexity, 111–114, 132–136
Stressors
  neural, 54
  systemic, 54
Sucking, 20, 21
Symbolic behavior, 209
Symbolic representation, 9, 146, 148, 153, 225
Symbolism, 207
Syntax, 159, 166, 167, 174, 185
Systemic circulation, 54
Systemogenesis, 17, 31

T

Temperature regulation
  in babies, 71
  in rats, 18, 27, 55, 67
Tertiary circular responses, 110
Testosterone, 4, 48, 50
Therapy, 11, 217
Thresholds of auditory stimulation, 101
Timidity, 42, 63
Trial and error learning, 15, 30

V

Vocabulary, 10, 159, 170, 198, 204
Visual fixation time, 109

W

Weaning in rats, 3, 5, 16, 18, 24, 32, 35, 39, 43